THE HUMAN BODY IN MINUTES

TOM JACKSON

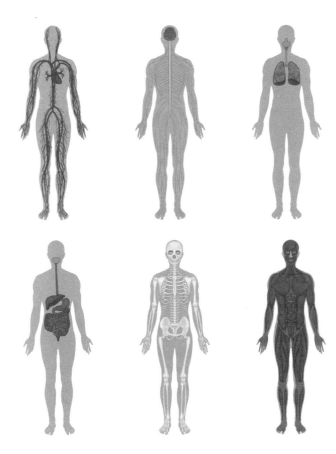

THE HUMAN BODY IN MINUTES

TOM JACKSON

Quercus

CONTENTS

Introduction   6
Your body   8
Support and motion   30
The digestive system   64
The respiratory system   88
The cardiovascular system   102
The circulatory system   114
The immune system   122
The nervous system   140
The skin   180
Bodily secretions   202

Homeostasis	224
The reproductive system	238
Inheritance	284
Human behaviour	302
Human evolution	328
Diet and nutrition	348
Disease and medicine	364
The future of the body	398
Glossary	408
Index	412
Acknowledgements	416

Introduction

The human body is a subject of endless fascination; no other living thing has been investigated so thoroughly and in so many different ways. On one level, your body is a biological entity made up of some 30 trillion cells that live and die together. On another, it can be treated as a collection of body systems, performing functions that support each other and maintain overall health and wellbeing (or at least, can be helped to do so with medical attention). What's more, the body is also the vehicle for a person's identity – what are you without it?

Our understanding of the human body draws on a great many fields of scientific research – biochemistry, microscopy, psychology and genetics to name just a few. Medical science seeks to understand how the body works so that we can know how best to intervene when things go wrong due to illness or injury. To do that every aspect of the body must be studied, from the individual cells and tissues to the large-scale or gross anatomy of the organs and limbs.

We can answer many questions. How do bones and muscles create movement, and what are the mechanisms that orchestrate it all so seamlessly? How does the food we eat keep us alive, and how do nerves send electrical signals around the body? Why do we hiccup, get old, become ill and recover? How does a new human body get made, and what can the study of ancient human evolution reveal to us about why our bodies work the way they do today?

However, there is more. As complex biological entities, human bodies adopt particular behaviours that extend beyond the merely physical. We gather together in different social groups, communicate and cooperate with each other. Of particular importance is our consciousness – the internal mental life seated in the human brain. How does our obvious and startling intellect arise from a biology that we largely share with other animals, and can we be sure we are unique in this respect?

In its exploration of the human body, the goal of this book is to take a good look at ourselves. There is much to learn along the way, but our bodies still pose questions that remain unanswered and, as we shall see, there are many possibilities for the future of the human body.

Your body

All at once the human body can appear mundane and incredible. Despite much fretting over its precise shape and size, the variation between two bodies is contained within a small range of differences. However, when we consider the feats of metabolism and physiology that our bodies perform second by second for decades, one cannot help but be amazed by this slick biological machine, more intricate than any constructed device.

Of course, we are not alone in having such a body. Animals of all stripes (and spots and tentacles) share many of the same features, and outdo us with many others. Nevertheless, we humans have some unique features that set us apart. First, we are bipedal, meaning we walk on two legs. No other mammal habitually moves like this, and our upper limbs are used instead for carrying and manipulating objects. That ability goes hand-in-hand with a second defining feature, the human brain, said (by humans) to be the most complex system in the Universe.

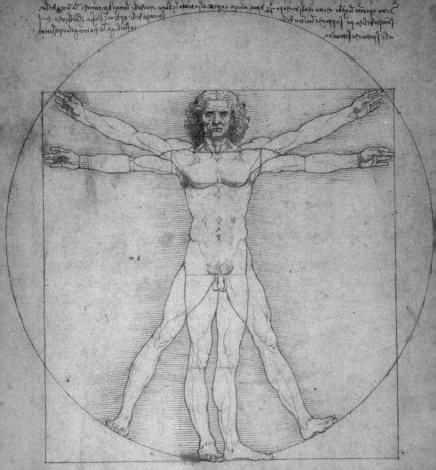

The average woman

The average female human is about 15 per cent smaller than the average male human, by weight. There is a slightly lower chance of her being born, with 48.1 per cent of all births being girls. However, she is likely to live longer than her male counterpart. The global average life expectancy of a woman is currently 73 years and 6 months, although this varies widely from nearly 87 in Japan to less than 51 in Sierra Leone.

The tallest women live in Latvia, where the average female height is 170 centimetres (5 ft 6 in). By contrast the average is 149 centimetres (4 ft 10 in) in Guatemala. Average weights follow a different pattern, with women in Bangladesh averaging a 19.8 body mass index (BMI; see page 362), while those living in Kuwait average a BMI of 31.4.

The average woman gives birth 2.45 times, markedly down from five births in the 1960s. South Korean women have the fewest children (1.2 births), while in Niger the average is 7.6.

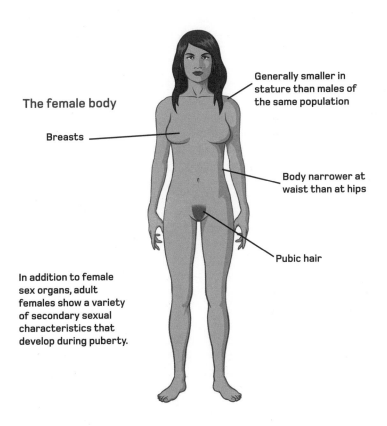

The female body

Breasts

Generally smaller in stature than males of the same population

Body narrower at waist than at hips

Pubic hair

In addition to female sex organs, adult females show a variety of secondary sexual characteristics that develop during puberty.

The average man

The male is the largest and strongest of the human sexes, but life is, on average, shorter and more dangerous for males than for females. For every 100 girls born, approximately 108 boys are born. However, by the age of 65 the ratio has flipped to approximately 0.75 males for every female, with considerable variation between cultures. This is partly due to males being more susceptible to particular diseases, such as haemophilia, but the change is also due to risky and violent behaviours (and war), which dramatically increase the male death rate after the age of 15. The average life expectancy of a man is 68 and 6 months. In Switzerland that goes above 81, while in Sierra Leone it is 48 and 4 months.

The tallest men are from the Netherlands, who have an average height of 183 centimetres (6 ft), while the shortest live in East Timor where the average is 160 centimetres (5 ft 2 in). The thinnest men are from Eritrea, with a BMI of 20.1, while Argentinian men have the highest BMI, at an average of 28.7.

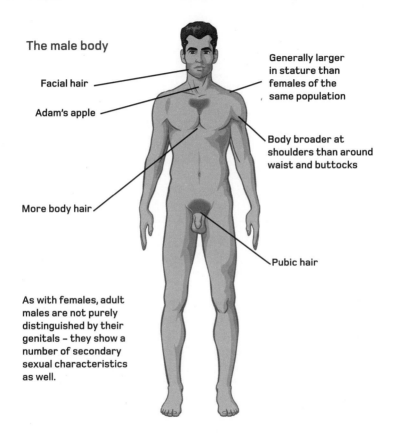

The male body

Facial hair

Adam's apple

More body hair

Generally larger in stature than females of the same population

Body broader at shoulders than around waist and buttocks

Pubic hair

As with females, adult males are not purely distinguished by their genitals – they show a number of secondary sexual characteristics as well.

What is the body made of?

Until the mid-19th century, it was thought that the human body, along with other living entities, was put together by the action of a mysterious, and probably divine, 'vital force'. This force separated the animate from the inanimate, and the ordinary processes of physical chemistry did not apply to the organic workings of the body. Then in 1828, Friedrich Wöhler produced urea – an organic chemical found in urine – in the laboratory completely by accident. This showed the world that the chemistry of life was no different to the chemistry of everything else, and the field of biochemistry was born.

Analysis of the human body showed that 99 per cent of its mass was made of six elements: oxygen, carbon, hydrogen, nitrogen, calcium and phosphorus (60 per cent of the mass is water, a compound of hydrogen and oxygen). Another 56 elements are found in tiny amounts, including iron, beryllium, arsenic and radium. Of these, 15 are deemed essential (see page 354). The others may have as-yet unknown functions, or may be natural impurities.

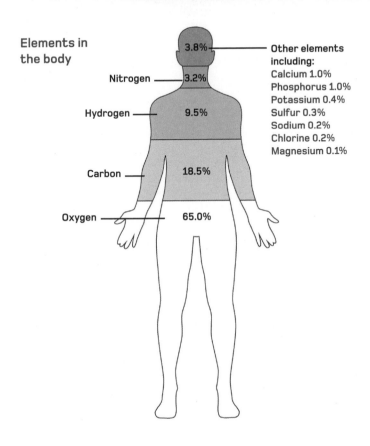

Elements in the body

Other elements including:
Calcium 1.0%
Phosphorus 1.0%
Potassium 0.4%
Sulfur 0.3%
Sodium 0.2%
Chlorine 0.2%
Magnesium 0.1%

3.8%

Nitrogen — 3.2%

Hydrogen — 9.5%

Carbon — 18.5%

Oxygen — 65.0%

Cells

The human body is built from structural units called cells. It is estimated that the average body contains 37 trillion of them. Making that judgement is far from easy, because body cells range in size from 4 micrometres for a granule cell in the cerebellum to 1 metre (39 in) for a motor neurone in the leg. However, human body cells still share many features.

The cell is surrounded by a membrane. This is flexible, and so the shape of the cell can change. While the membrane encloses the cell, substances can move through it in both directions, mostly via pores in its surface. Inside, the cell is filled with cytoplasm, a watery mixture of hundreds of dissolved chemicals. There is also a nucleus, which is surrounded by its own membrane. This is where the cell's DNA (see page 286) is stored. Under the control of that DNA, the cell's myriad chemicals are manufactured in highly folded membranous structures called the endoplasmic reticulum. Mitochondria power the cell's activities by releasing useful energy from sugar fuel.

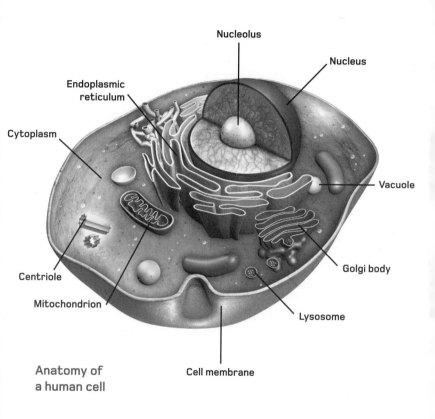

Nucleolus

Nucleus

Endoplasmic reticulum

Cytoplasm

Vacuole

Centriole

Golgi body

Mitochondrion

Lysosome

Cell membrane

Anatomy of
a human cell

Metabolism

Metabolism is the catch-all term for the chemical activity of a body. In the round, metabolism processes food and other chemicals entering the body to provide the energy for life and to help grow and maintain the body. The idea can be divided into anabolism and catabolism. The former means 'to build up' and the latter means 'to break down'.

Anabolism involves taking simple, raw materials and making them into more complex and useful substances. That could be chaining amino acids into a protein. By contrast, catabolism would take a protein from food and break it into its constituents for later use. Both processes combine to form metabolic pathways that involve dozens of chemical steps heading towards a specific product – an enzyme, a hormone or a fat molecule, for example. There are 260 major pathways at work in the human body. Most steps in these pathways use energy and, therefore, perhaps the most crucial human pathway of all is glycolysis, which kick-starts the release of a life-sustaining supply of energy from sugars.

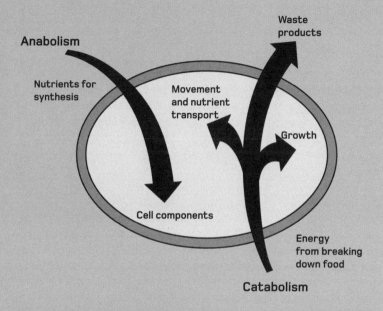

Anabolism

Nutrients for synthesis

Movement and nutrient transport

Waste products

Growth

Cell components

Energy from breaking down food

Catabolism

Basic metabolic processes

Respiration

For many, the word 'respiration' means breathing, but the term also relates to the metabolic pathway that extracts the energy from sugar fuel to power the body. Respiration takes place in a cell's mitochondria, where fuel in the form of glucose is oxidized. This turns each glucose molecule into six carbon dioxide and six water molecules, with a release of energy in the process.

This reaction is the same as igniting the sugar, but the cell does it in a more gradual way. First, a 10-step pathway called glycolysis converts glucose into the chemical pyruvate, releasing energy. Pyruvate then starts the next phase, the Krebs cycle, which has 10 steps that result in water and carbon dioxide – and more energy. The cell makes use of energy carriers such as adenosine triphosphate (ATP). To transfer its energy to a metabolic reaction, ATP ejects a phosphate and forms diphosphate (ADP). During respiration ADPs and similar energy carriers receive the energy required to recharge, collect a third phosphate and form ATP again. Each glucose molecule will recharge 32 ATPs.

$$C_6H_{12}O_6 + 6O_2 \rightarrow 6CO_2 + 6H_2O + energy$$

glucose + oxygen $\rightarrow$ carbon dioxide + water + energy

Chemicals for life

The human body lives and dies by the supply of essential chemicals that are used as fuel and building materials. In the simplest terms, the human body is built from three materials: carbohydrate, protein and fat.

Carbohydrates are compounds based on sugars such as glucose. In the human body, sugars are the primary source of energy and for storage – mostly in the liver or muscle cells – they are bundled into a complex form called glycogen. Sugars also have a structural use when linked to proteins. Proteins are the dominant material in muscles and structural elements of the skin. They are also used as enzymes, the body's worker chemicals. All the steps in a metabolic process are facilitated by an enzyme holding and manipulating the chemicals involved. Lastly, fats are the main material in cell membranes. Fatty sheaths also coat major nerves, while oils keep skin and hair healthy and, of course, excess food intake is converted to fat as the body's long-term energy store.

Food group	Food sources	Key functions
Carbohydrates (sugars, starch, cellulose etc.)	Rice, potatoes, cereals, fruit, vegetables	Energy for rapid use, synthesis of structural materials such as glycoproteins, digestive fibre
Proteins (complex molecules formed from chains of amino-acid units)	Meat, fish, dairy products, eggs, soy, grains and cereals, legumes, nuts and fruit	Raw materials for synthesis of most body tissues, also an energy source
Fats (compounds of 'fatty acids' with glycerol)	Fatty meats, dairy products, plant oils and nut oils	Energy for immediate use, long-term energy storage, transport of vitamins A, D, E and K in the body

Tissues

There are about 200 different cell types in the human body, each one built to do a particular job: the red blood cell carries oxygen around the body; the nerve cell transmits signals; the goblet cell in your nose lining produces mucus. Biologists like to keep things organized, and since no cell works alone they are grouped into tissues, or a collection of similar cells that work together to perform a particular role in the body.

Of the four broad types of tissue in the human body, a group of muscle cells makes muscle tissue and the brain is built from nervous tissue. There are also connective and epithelial tissues. Connective tissues make up the structural parts of the body, and so include bone and cartilage (like ligaments or the outer ear) and also the blood vessels and blood cells. Epithelial tissues cover the surface of a body. They include hair and the outer layer of the skin, but also the linings of cavities within the body, such as the mouth and gut, the airways and the reproductive tracts.

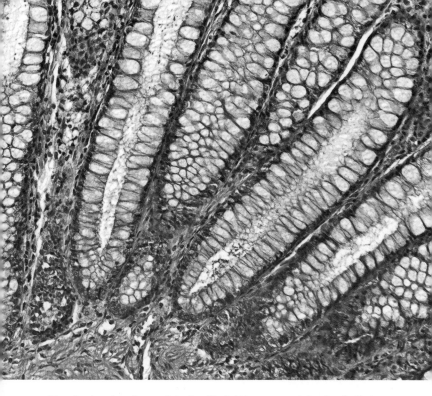

The glands primarily consist of epithelial tissue – specialized cells that secrete substances such as saliva, mucus and hormones.

Organs

The human body is a highly regimented system with ranks and hierarchies. The lowliest unit is the cell, which forms groups of tissue. Tissues seldom work alone, but require input from other body parts, including other tissues. A collection of tissues that work together to fulfil a function is called an organ.

There are some obvious examples – the brain and the heart – but the term can be used for more lowly structures. The ear, nose and testicle are all organs, for example. While undesirable, it is possible for the body to survive without one or more of these organs. But it could not withstand the loss of one of the so-called vital organs: brain, heart, liver, kidneys, lungs, stomach and intestines.

The skin is also an organ – much overlooked in these lists – and one that performs a large number of important jobs. Other organs include the pancreas, the spleen and the bladder, while a pregnant mother develops a temporary organ, the placenta, as a conduit between her body and the developing fetus.

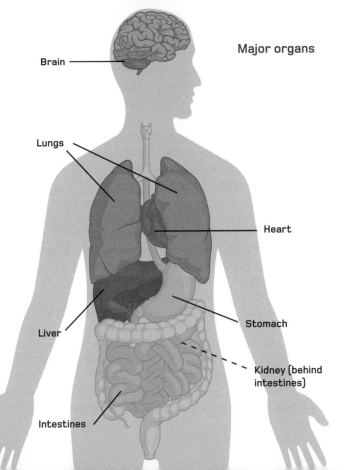

Major organs

Brain

Lungs

Heart

Liver

Stomach

Kidney (behind intestines)

Intestines

Body systems

The organs are the largest single structures in the body, but we get a better understanding of what they do by grouping them into body systems. This method of organization is an artificial one, and there are a number of ways it can be done.

The circulatory system comprises the heart and blood vessels. It acts as a transport network for the body, and connects to the respiratory and digestive systems. The former includes the lungs and airways used to bring oxygen into the body and expel carbon dioxide. The latter is the food-management system and includes the mouth, stomach and intestines, but also the liver, which processes nutrients before they enter the bloodstream. Other body systems include the nervous system, the musculoskeletal system (often split in two), the reproductive system and the integumentary system (skin, hair, nails). There are also the lymphatic and excretory systems, which remove unwanted materials, the endocrine system, made up of hormone glands, and the immune system, which defends the body against attack.

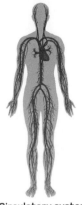

Circulatory system

Nervous system

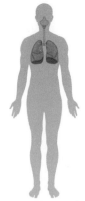

Respiratory system

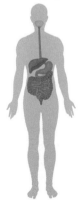

Digestive system

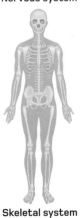

Skeletal system

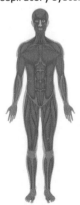

Muscular system

Support and motion

Like all vertebrates, humans have an internal skeleton that forms the rigid structure underlying their body shape. This skeletal system is made up of bones and cartilaginous features, such as ligaments and tendons.

The skeletal system's role is not just to create a solid shape for the body. It also provides anchor points for muscles that are then able to alter the body shape by bending it at joints. There are 340 joints in the body, 230 of which have at least one degree of freedom, meaning they can move around one axis only. Some, such as the hip or neck, are able to move in several directions. This means that the human skeleton can move in 244 independent ways.

The main purpose of altering the body shape is for locomotion. The human body is adapted best for walking, but it can also run, jump, climb and swim. Another significant motion in humans is movement of the arm and hand for manipulating objects.

Major muscles

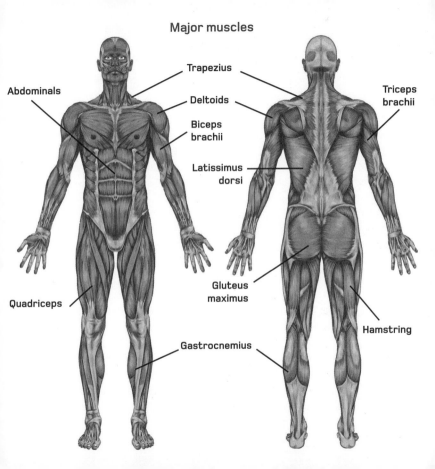

Trapezius

Abdominals

Deltoids

Biceps brachii

Triceps brachii

Latissimus dorsi

Quadriceps

Gluteus maximus

Hamstring

Gastrocnemius

The skeleton

The adult human skeleton contains 206 bones. At birth that number is 270, but as the body grows and develops many of these fuse together. The largest bone in the skeleton is the femur, which connects the hip to the knee. The smallest is the stapes, or stirrup. This is about 3 millimetres long and is a non-structural ossicle, or ear bone, used to transmit sound.

The skeleton is divided into two parts. 80 bones comprising roughly the skull, spine and ribs form the axial skeleton. The remaining number make up the appendicular skeleton, which, as the name suggests, includes the appendages. At the top, the shoulder girdle connects to the arm bones, while the pelvic girdle joins to the legs. Further adjectives help describe various positions and directions in the skeleton. Towards the head is cranial; the rear end is caudal. The front, or chest side, is ventral; the back is dorsal. Structures at the extremities are distal; those near the core are proximal. Finally, a movement away from the body is lateral, while towards it is termed medial.

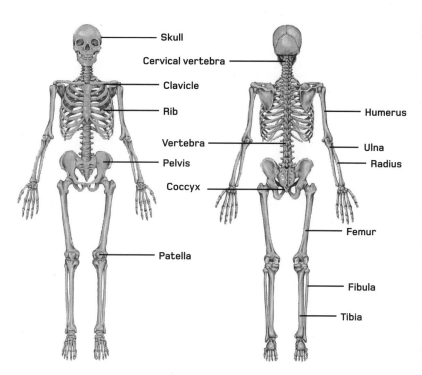

The human skeleton

Bone

Around 15 per cent of human body mass is bone. Besides its structural role, bone also acts as a mineral storage system and is where new blood cells are manufactured. Bone has three distinct parts: compact bone, spongy bone and bone marrow.

Compact bone forms the hard, dense outer layer. Its stiffness comes from the mineral hydroxylapatite, a hydrated form of calcium phosphate. This mineral creates a solid matrix around bone cells, or osteoblasts, from a protein mixture called osteoid, which is secreted by the cells. Compact bone is very much alive; osteoblasts and their surrounding matrix are continuously replaced as other cells, called osteoclasts, break down and reabsorb older parts of the bone. Beneath the compact layer is spongy bone. Here, the matrix becomes porous and filled with spaces and so has a very large surface area. The spaces are filled with bone marrow. This is where new red and white blood cells are produced – mostly in the heads of larger bones. Each day a human skeleton produces 500 billion blood cells.

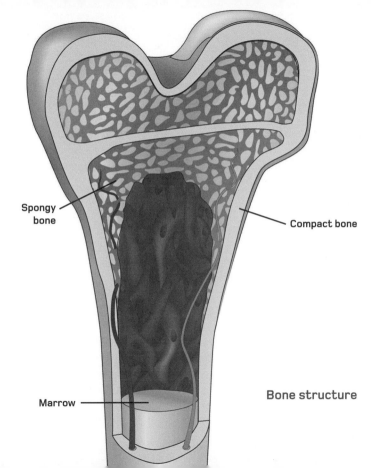

Spongy bone

Compact bone

Marrow

Bone structure

Cartilage

The external anatomy of the human body owes much to a connective tissue called cartilage. At birth, most of the skeleton is made of cartilage, which transforms into bones over the first six years or so of life.

Cartilage is an elastic substance that is more flexible than bone, but more rigid than muscle. Like bone, it is not a mass of inanimate material, but is composed of cells, called chondrocytes. These secrete a protein mixture that forms a solid matrix around them. The matrix is made up largely of collagen – a tough fibrous substance that is the main structural protein in the body – and elastin, which provides a degree of elasticity. The proportions of these vary according to function. Tough hyaline cartilage is used to create a cushion between bones at joints. It is also in the trachea (windpipe), larynx (voice box), between the ribs and makes most of the nose. Elastic cartilage makes more flexible structures like the outer ear and epiglottis. Fibrocartilage is found in the most flexible joints like the knee, wrist and shoulder.

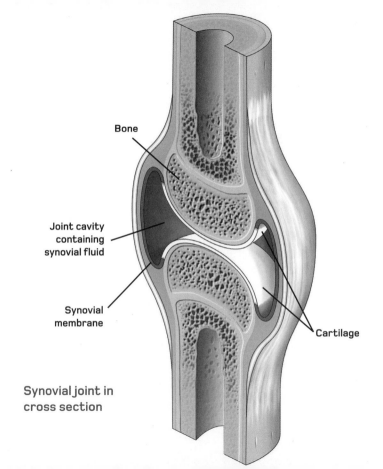

Bone

Joint cavity containing synovial fluid

Synovial membrane

Cartilage

Synovial joint in cross section

Ligaments and tendons

The skeletal and muscular systems are held together by connective fibres called ligaments and tendons. Ligaments connect bone to bone and maintain the shape and degree of movement at joints. For example, a dislocated bone is one that is no longer held in place by its ligaments. Tendons connect muscles to bones. They are stiffer and less elastic than ligaments so the contraction of a muscle is transmitted fully to the bone, making it move. Both ligaments and tendons are examples of fibrous connective tissue, made up mostly of collagen. There is a third member of the set – fascia. Instead of being cord-like, this is a sheet of collagen that holds muscle groups together and separates them from internal organs.

Damage to ligaments and tendons can be permanent. They are able to stretch a little for a short time and then spring back to their original form, but if they are extended for long periods, such as after an injury, they begin to lose this elasticity. Further problems are likely to occur if not treated quickly.

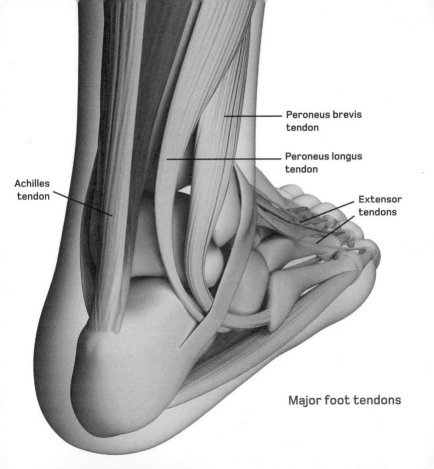

Peroneus brevis
tendon

Peroneus longus
tendon

Achilles
tendon

Extensor
tendons

Major foot tendons

Joints

A joint is any connection between bones. There are 340 joints in the human body, since many of the 206 bones are connected to more than one of the others. Some joints have bones connected by cartilage, and these have very limited ranges of motion. A suture is where bone is knitted to bone with collagen, and these are more or less immobile. What most of us think of as joints are the articulated regions of the skeleton, known as synovial joints. Here the bones do not touch but are connected by a fluid-filled capsule made from cartilage.

There are six types of synovial joint: a hinge allows movement in just one plane (elbow). A pivot offers a circular, twisting motion (forearm). An ellipsoidal joint allows movement in all directions (base of the fingers), while a ball-and-socket joint is similar and allows for even larger movements (shoulder). A gliding joint, found in the wrist, allows bones to move a little from side to side without rotating. Finally, the saddle joint, which is found in the thumb, allows both a back-and-forth and side-to-side motion.

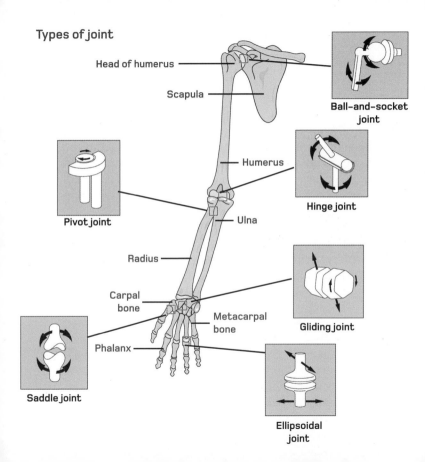

Types of joint

Head of humerus

Scapula

Ball-and-socket joint

Humerus

Pivot joint

Hinge joint

Ulna

Radius

Carpal bone

Metacarpal bone

Gliding joint

Phalanx

Saddle joint

Ellipsoidal joint

The skull

Of the 29 bones that make up the human head, the heaviest is the mandible or jawbone. Eight more bones form the rounded cranium and 14 are facial bones. Then there are six tiny ossicles – or ear bones – three each side (see page 170). The teeth, fixed to the mandible below and the maxilla bone above, are part of the skull but are not made from bone.

The mandible is the only skull bone that is articulated. It forms a joint with the temporal bones on each side of the cranium, so it can open and close the mouth. All other cranial or facial bones are connected by immobile sutures, forming a cohesive case.

Newborns have 44 bones in their skulls, with six gaps called fontanelles. These close up as the bones fuse into the adult anatomy. Even in adulthood the skull retains 22 holes, or foramina. Most are small and accommodate cranial nerves and blood vessels. At the base of the skull is the foramen magnum, through which the spinal cord connects to the hindbrain.

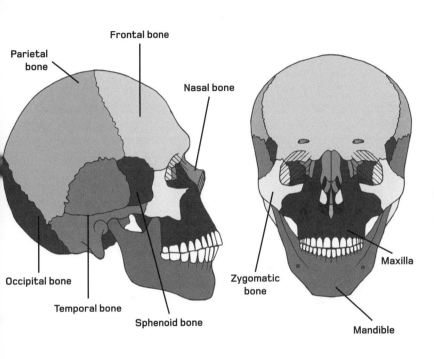

Bones of the skull

The spine

The spine is made up of a column of bones called vertebrae. Most people have 33, although a 32-bone spine is not uncommon. The vertebrae are connected by cartilage – thus each one has a limited range of movement – but together they create a flexible unit at the heart of the axial skeleton.

Inherited from our four-legged cousins, the spinal column has rotated to a vertical position, but retained a curved, double-bow shape. This active posture is the primary cause of back trouble in our inactive society. The spine supports the weight of the head, about 11 pounds (5 kg), and protects the spinal cord inside a canal that runs through the centre of each vertebra. The column has five regions. The cervical spine forms the neck; the thoracic spine supports the ribcage; and the lumbar spine in the lower back contains the largest vertebrae. The end of the spine is made up of the sacrum – five vertebrae fused into one. The pelvis connects here. The very tip is the coccyx. This single bone made of fused sections is a vestigial remnant of the mammalian tail.

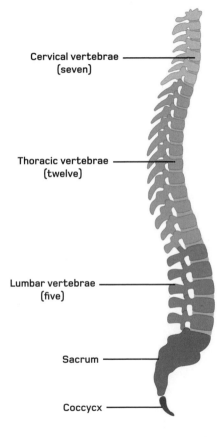

The vertebral
column

Cervical vertebrae
(seven)

Thoracic vertebrae
(twelve)

Lumbar vertebrae
(five)

Sacrum

Coccycx

The thorax

This region of the body, between the head and the belly, is a cage of bone that protects the most vital of organs – the heart and lungs – and the massive blood vessels that connect them and the rest of the body.

This cage is composed of 12 pairs of curved rib bones that protrude from the vertebral column at the back. The bones do not continue all the way to the front. Instead, the final section is costal cartilage, which connects to the sternum, or breastbone. The first seven ribs connect directly to the sternum. The next three – so-called false ribs – share their cartilaginous connection, while the final two are floating ribs in that they are not connected to anything.

As well as providing protection, the ribs are able to move and are part of the process of breathing. Intercostal muscles run between the ribs. As these contract they expand the ribcage, helping to draw air into the lungs.

The ribcage

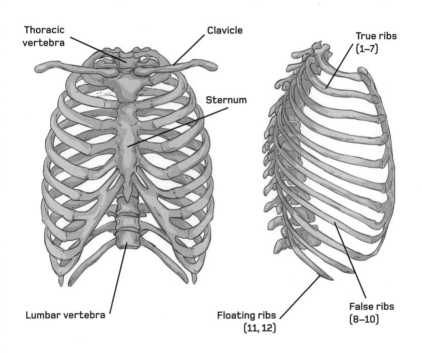

Thoracic vertebra

Clavicle

True ribs (1–7)

Sternum

False ribs (8–10)

Floating ribs (11, 12)

Lumbar vertebra

The pelvis

The pelvis is a girdle of bones around the lower abdomen, which forms the connection between the spine and the lower limbs. It also houses the posterior orifices, or body openings – namely, the anus, urethral orifice (or urinary meatus) from which urine is discharged and, in the case of females, the vagina or birth canal. To make room for childbirth, the female pelvis is wider and rounder than that of the male.

Unlike the skeleton's other girdle – the shoulder, or pectoral, girdle – the pelvis is relatively rigid. It is made of two curved hip bones, which swing around from the sacrum at the base of the spine. Each hip bone is made of three bones that become fused during puberty. The largest is the ilium, which rises above the girdle, while the curved ischium loops below it. The pubis lies between the two. The girdle is completed at the front with cartilage knitting the two hip bones together. The head of the femur fits into a socket where these three bones connect, creating the ball-and-socket hip joint.

The pelvic girdle

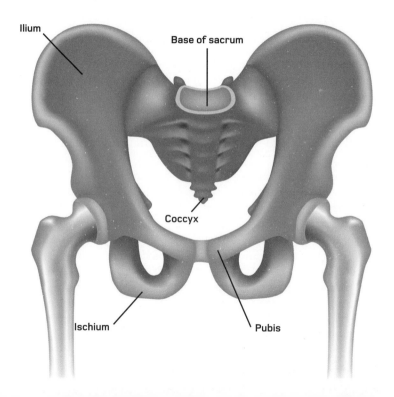

Ilium

Base of sacrum

Coccyx

Ischium

Pubis

Limbs

The arms and legs are the main components of the appendicular skeleton, connecting to the axial skeleton via bone girdles. The pelvic girdle is a rigid anchor for the legs, while the pectoral girdle, composed of the scapula (shoulder blade) and clavicle (collar bone), is considerably less substantial, affording the arms an enormous range of movement.

In medical vernacular, the term 'arm' only refers to the upper region above the elbow. The rest is the forearm. Conversely 'leg' refers to the region below the knee, the part above being the thigh. Both upper and lower limbs have an analogous arrangement of bones. The thigh and arm have one long bone – the femur and humerus (funny bone), respectively – while both the leg and forearm have two bones of unequal size. Both limbs also have analogous hinge joints, in the elbow and the knee. The elbow has a greater range of movement, allowing the hand to move towards and away from the body. The knee is altogether sturdier and allows for more rotation to aid with walking.

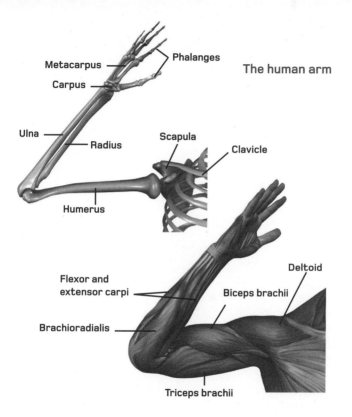

The human arm

Metacarpus

Phalanges

Carpus

Ulna

Radius

Scapula

Clavicle

Humerus

Flexor and
extensor carpi

Deltoid

Biceps brachii

Brachioradialis

Triceps brachii

The hand

We humans use our hands to hold and manipulate objects and to investigate the world. Only the tongue rivals the hand in its density of touch sensors, and no other body part has such an intricate description, divided as it is into palm, fist, heel, fingertips and knuckles, not to mention the five individually named digits – the four fingers and thumb.

The grasping nature of the hand is achieved via 27 bones and 34 muscles. Each finger has three articulated phalanges, the thumb has two, but makes up for this by using one of the five metacarpals; the other four form the palm. The wrist contains eight carpals. The hand closes into a fist using muscles on the underside of the forearm. It straightens again using muscles on the back of the hand. Although it is not considered a true finger, the thumb is the dominant digit. It can bend like the others, but also rotate in all directions and, most significantly, can move in the opposite direction to the fingers. This ability creates both a strong grip and fine motor skills.

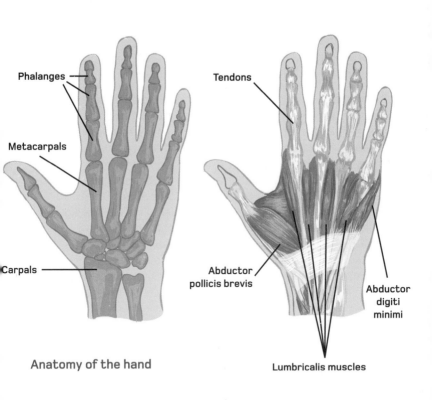

Phalanges

Metacarpals

Carpals

Tendons

Abductor
pollicis brevis

Abductor
digiti
minimi

Lumbricalis muscles

Anatomy of the hand

The foot

The foot is the only body part habitually in contact with the ground. As the leg is to the arm, the foot's structure is analogous to that of the hand, although there are differences that allow the foot to carry and balance the body's weight.

The five digits, or toes, each have three phalanges, except the hallux (big toe), which has two. The sole of the foot is formed of five metatarsals, which are generally longer than the digits – unlike in the hand. As a result, the foot is considerably less flexible, despite its 20 articulated joints and dozens of muscles. The ankle joint, or tarsus, is the most flexible part of the foot and comprises seven bones that are the foot's equivalent of the carpals. The lower leg bones connect to the talus (ankle bone) and weight is transmitted into the calcaneus (heel bone), which is the largest bone in the foot. The tarsus also includes five other bones, which protrude forward to make the arch of the foot. This rigid feature works like a spring, flexing slightly as weight is applied and released during running and walking.

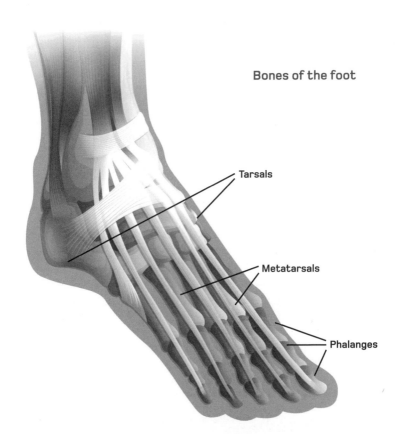

Bones of the foot

Tarsals

Metatarsals

Phalanges

Muscles

There are three types of muscle in the human body: skeletal, smooth and cardiac. The 650 skeletal muscles are attached to the bones. Some are used to create voluntary movements, while others work involuntarily (beyond our control), to maintain posture and balance.

Smooth muscles are found in the digestive system, the airways, blood vessels and reproductive organs. They attach to the internal organs and skin and their action is involuntary. Cardiac muscle is found in the heart. It does not tire in the way that other muscles do and, again, its action is involuntary. All three muscle types work in the same fundamental way: they apply a force by contracting or reducing in length. There are some major differences, however. Skeletal muscle appears striped or striated due to the dense packing of the contractile proteins. Its individual cells are also merged into long fibres. Cardiac muscle is also striated but maintains individual cells. Smooth muscle cells are less dense and do not show the striations.

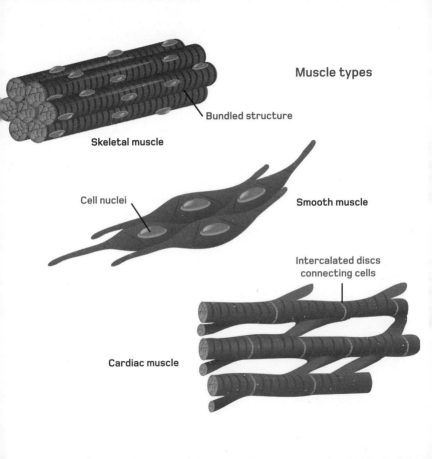

Muscle types

Bundled structure

Skeletal muscle

Cell nuclei

Smooth muscle

Intercalated discs
connecting cells

Cardiac muscle

Contraction

A change in the shape of a contracting muscle creates a force that moves a body part. This physical change is produced by a chemical change, specifically in two proteins found in muscle cells: actin and myosin. These long protein molecules are arranged alternately in layers. In striated muscle they are bundled into dense units called sarcomeres, which are orientated the same way to create a cumulative effect. In skeletal muscle, hundreds of thousands line up to make fibres called myofibrils, coated with mitochondria to provide power.

Contraction begins with the release of a neurotransmitter from a nerve. This stimulates a wave of electricity along the muscle fibre. This causes the slender actin molecules to haul themselves along the thicker myosin ones. Each sarcomere is just two-millionths of a metre long, and contracts by about 10 per cent. However, due to the specific arrangement of millions of fibres, some muscles contract by more than half their length.

Muscle structure

Contractile muscle

Myofibril
(individual muscle thread)

Sarcomere
(myofibril subunit)

Thick
filament
(myosin)

Thin filaments
(actin)

Antagonistic pairings

A muscle never works alone. It can only pull on a bone, or other body part, but never push. It therefore relies on one or more partners to produce the opposite motion. This system of musculature is called antagonistic pairing.

The most familiar example of this system is the biceps and triceps pairing in the arm, which articulates the elbow. The biceps, named for its two connects to the scapula (shoulder blade), is the flexor muscle. This means it bends the elbow. At its distal end, the biceps attaches to the radius in the forearm. As the muscle contracts, the forearm is levered towards the body, with the elbow joint acting as the pivot. To straighten the arm, it is not enough simply to relax the biceps. The connective tissue around the joint maintains the position. So the extensor muscle takes over. This is the triceps (a three-headed muscle connected to the scapula and humerus), located on the reverse side of the arm to the biceps. As this contracts, it pulls on the ulna (next to the radius), making the arm straighten.

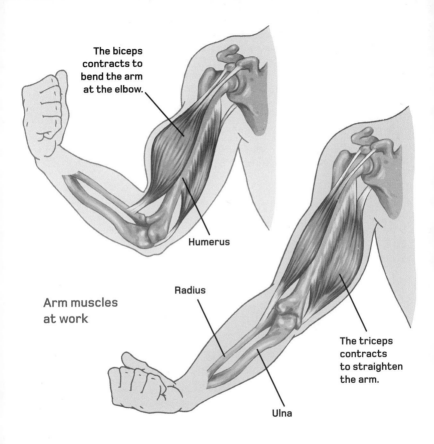

The biceps contracts to bend the arm at the elbow.

Humerus

Radius

Arm muscles at work

Ulna

The triceps contracts to straighten the arm.

Walking and running

A human can move in several ways, but the most significant gait is walking. A runner reaches top speed after about 10 seconds and then begins to slow owing to fatigue after another 10. However, the average human can walk for several hours, covering 5 kilometres (3 miles) per hour.

Walking is more efficient than running. The latter involves the whole body leaving the ground, by making forward leaps with every stride. In contrast, when walking, one foot is always in contact with the ground and there is a barely perceptible rise and fall of the body. Each leg works as a double pendulum. The first pendulum is the leading leg swinging forward from the hip. The body weight shifts forward as the front foot hits the ground, heel first. The second pendulum motion is from the roll of the foot, which shifts weight onto the toes, and allows the other leg to begin its pendulum swing. About 60 per cent of the motion energy used is recovered due to the springiness of the feet and legs.

Anatomy of a walking man

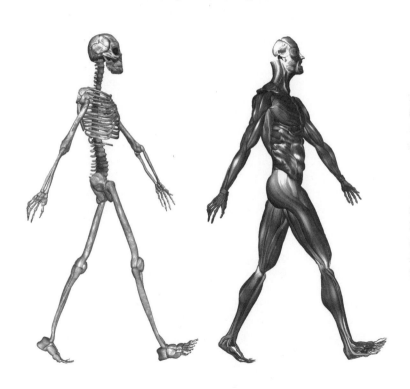

The digestive system

All of the nutrition and hydration – food and water – that enters the body arrives through the digestive system. It is here that digestion, absorption and excretion take place – a series of physical and chemical processes that break food down into simpler constituents, in order to absorb the useful nutrients within it and then to expel any unwanted materials.

The digestive system is based around the alimentary canal, a convoluted tube that begins at the mouth and ends at the anus. Along the way it includes the oesophagus, stomach, intestines and rectum. The canal is made up of connective tissues – mostly smooth muscle – and then lined with epithelial tissue. Epithelial tissue lines the surface of the body, and we find it in the digestive system because the alimentary canal is an exterior surface like the skin, albeit one that passes through the body core. The average alimentary canal is 9 metres (29½ ft) long. A solid piece of food in the system is known as a bolus. Once liquefied, the content is called chyme.

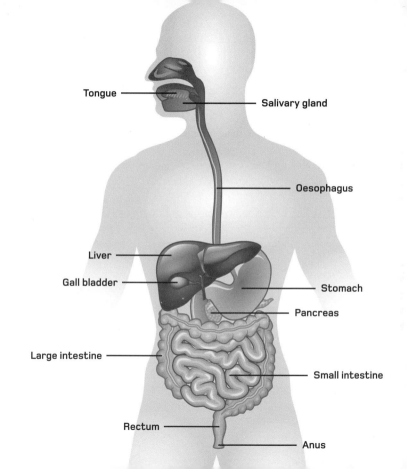

Tongue

Salivary gland

Oesophagus

Liver

Gall bladder

Stomach

Pancreas

Large intestine

Small intestine

Rectum

Anus

Hunger and satiety

The need to eat, to replenish supplies of energy and other nutrients, is a motivation we share with all animals. Humans experience this as the physical sensation of hunger. After about 12 hours without food, the empty stomach begins to contract, creating mild pains – the signal that the body is in need of more food. Satisfying this need results in a feeling of fullness (satiety).

Fortunately for most people today, extreme hunger is a rare experience. Nevertheless, we all feel hungry from time to time and, surprisingly for such a basic body function, the mechanisms that control hunger are poorly understood. Alongside the physical aspect described above, there is also a chemical factor in the form of hormones called leptin and grehlin. Produced in fatty tissues and the stomach lining respectively, these hormones inhibit and initiate hunger according to our energy supplies. However, our appetites are not only linked to physiological needs; we often find ourselves eating too much – or too little – for both social and emotional reasons.

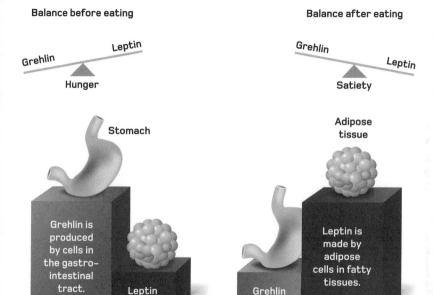

Grehlin and leptin are manufactured in different parts of the body at different times. Production of the 'hunger hormone' grehlin is inhibited when the stomach is stretched, while the 'satiety hormone' leptin is produced when adipose cells store energy after eating. Both hormones act on receptors in the hypothalamus.

Teeth

Once food enters the mouth, it is broken down physically by the teeth. Children are born without teeth, but soon start to grow a set of 20 deciduous (milk) teeth. At the age of about six, these teeth begin to fall out and adult teeth, which are large enough to fill a fully grown mouth, replace them.

There are 32 adult teeth, although it is quite common to have a few less or even a few more. They are rooted within the jaw and skull bones, the lower teeth mirrored by the set above. However, teeth are not made from bone. The outer surface is made from enamel, the hardest substance in the body, made from 96 per cent hydroxylapatite, the same calcium phosphate mineral in bone. Beneath that is a layer of dentin, a softer mixture of proteins and minerals. The centre of the tooth is called pulp. It contains the nerves and blood vessels. Humans have chisel-shaped incisors and pointed canines at the front. These overlap and bite with a scissor action. The premolars and molars at the back are flattened for grinding food.

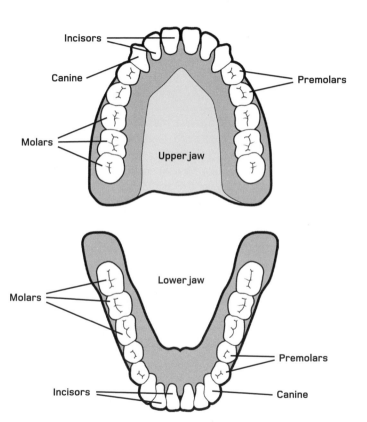

Saliva

Before food can be swallowed it must be converted into a slurry of liquids and small solid items. To this end, as the teeth chew, or masticate the food, it is mixed with a liquid called saliva. Saliva is produced by three main glands – under the tongue, in the lower jaw and in the throat. There are dozens more tiny glands in the gums and cheeks.

Saliva is produced all the time, but production spikes at the smell and taste of food. Made mostly of water, saliva contains mucin, one of the principle components of mucus. This is what gives saliva the slimy viscosity that protects the soft lining of the mouth and lubricates food prior to swallowing. It also contains enzymes – most notably amylase, which begins the digestion process even before the food leaves the mouth, by breaking down starch, a complex carbohydrate, into simpler sugars. Once the food is soft enough, it moves to the back of the mouth, where a complex wave of involuntary muscle contractions guide it into the oesophagus, while closing off access to the nose and windpipe.

Salivary glands and their ducts

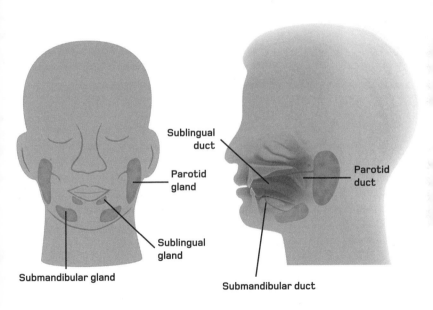

The stomach

The stomach is a muscular bag with an average capacity of 1.3 litres (2¼ pt). Its role is to mix food with enzymes and other active chemicals and churn them around for three to four hours to allow digestion to occur.

Food arrives from the mouth via the oesophagus. It enters through the cardiac sphincter, a ring of muscle that then closes tight to process the meal. At the other end of the stomach, the pyloric sphincter also stays shut to contain the contents until this phase of digestion is complete. Gastric juice contains dozens of active chemicals, which are secreted from the stomach lining. It has a pH of about 1.5 – somewhere between battery acid and lemon juice. This is due to the presence of hydrochloric acid. The stomach lining is protected by a layer of mucus, but if the juice leaks into the oesophagus, it creates the burning sensation of indigestion. The acid is there to promote the functions of enzymes, including pepsin, which breaks down proteins into smaller units called peptides, and lipase, which attacks lipids (fats).

Structure of the stomach

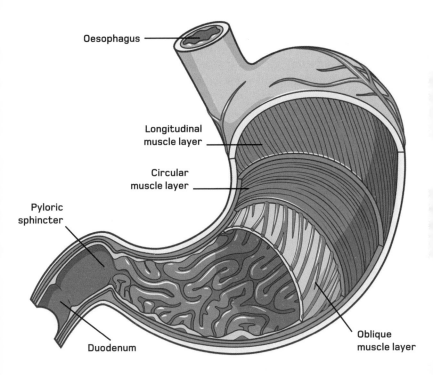

Oesophagus

Longitudinal muscle layer

Circular muscle layer

Pyloric sphincter

Duodenum

Oblique muscle layer

The gall bladder

After being processed in the stomach, a meal is then released through the lower sphincter into the duodenum, the short upper section of the small intestine. This is where the digestion process continues.

Trypsin arrives from the pancreas to break up peptides into amino acids, while lipases break down fat molecules into simpler fatty acids. However, lipids are immiscible with water – the major solvent throughout the digestive system – which means that the fats in food have the potential to clump. They resist chemical attack by the body, become impossible to absorb and simply pass through the system undigested. This is solved by adding bile, a salty and acidic yellow-green liquid that is made by the liver and delivered from a reservoir called the gall bladder. The bile duct pumps in about 40 millilitres (1½ fl oz) of bile into the duodenum. Salts in the bile surround droplets of lipids to form tiny packets called micelles. The micelles emulsify in the intestinal juices, and are now small enough to be absorbed through the intestine wall.

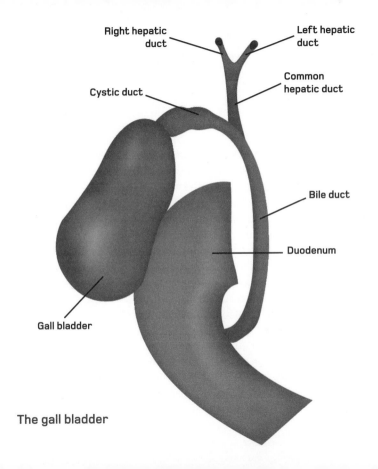

Right hepatic duct

Left hepatic duct

Common hepatic duct

Cystic duct

Bile duct

Duodenum

Gall bladder

The gall bladder

The small intestine

The small intestine is about 7 metres (23 ft) long, and generally longer in women than in men. That makes it considerably longer than the large intestine (the colon; see page 82), although at about 3 centimetres (1¼ in) across it is considerably narrower. The small intestine has three parts: duodenum, jejunum and ileum. The duodenum receives partially digested food, or chyme, from the stomach. It is the shortest section at around 25–35 centimetres (10–14 in), and its role is to digest the chyme into the smallest possible nutrients. It does this by receiving enzymes from the pancreas, the gall bladder and its own lining. The jejunum is about 2.5 metres (8 ft) long. Its primary role is to absorb the micronutrients – lipids, amino acids, sugars and vitamins – arriving from the duodenum. The ileum is the longest section, making up the remaining 4 metres (13 ft) or so. The ileum absorbs whatever nutrients are missed by the jejunum and also begins the job of retrieving enzymes, salts and other useful chemicals. The ileum then delivers a mixture of water and waste to the large intestine.

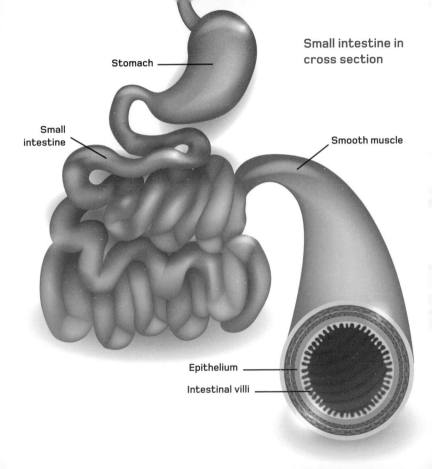

Stomach

Small intestine in cross section

Small intestine

Smooth muscle

Epithelium

Intestinal villi

Absorption of nutrients

The small intestine has an internal surface area of around 30 square metres (320 ft²), about the size of a space in a car park. This huge surface is required to absorb all the nutrients released by digestion, a process that takes about four hours for each meal. It is made possible by an intricate folding of the lining of the intestine, which creates finger-like projections called villi.

The wall of the small intestine

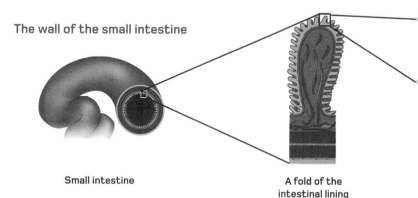

Small intestine

A fold of the
intestinal lining

Each villi is about a millimetre long and is lined with many epithelial cells, specifically column-shaped ones called enterocytes. To boost the absorption surface still further, the upper membranes of these cells are folded as well, creating a brush-like surface of microvilli around every villus. Enterocytes cannot rely on diffusion for nutrients to drift into them. They use a number of forms of active transport, using energy to pull material into the cells. They absorb salts and ions, vitamins, sugars, amino acids and fats. Most of these enter the blood vessels inside each villus and travel on to the liver. Insoluble fats, however, would block blood vessels and so enter the body via the lymphatic system.

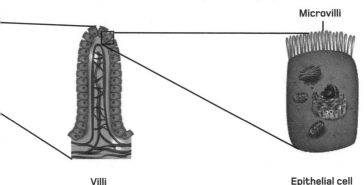

Microvilli

Villi
(c.1mm long)

Epithelial cell

The liver

The liver performs around 500 roles, most frequently forming the link between the digestive and circulatory systems. In some ways it is the body's largest gland, involved in manufacturing chemicals for the digestive and immune systems. It is also a store of glycogen, the body's quick-release energy supply.

The liver detoxifies various nutrients, most notably removing unwanted nitrogen compounds from proteins, which are eventually excreted in urine. It gets rid of the unwanted haemoglobin from old red blood cells by adding it to bile. This gives the bile its yellowish colour and ultimately makes faeces brown. The liver is also where nutrients absorbed in the intestine are processed before delivery to the rest of the body. The nutrients arrive from the intestine via the hepatic portal vein. This provides 75 per cent of the liver's blood, and it connects to thousands of functional units called lobules. The products of the liver mix with arterial blood arriving from the heart and then leave via the inferior vena cava, one of the largest veins in the body.

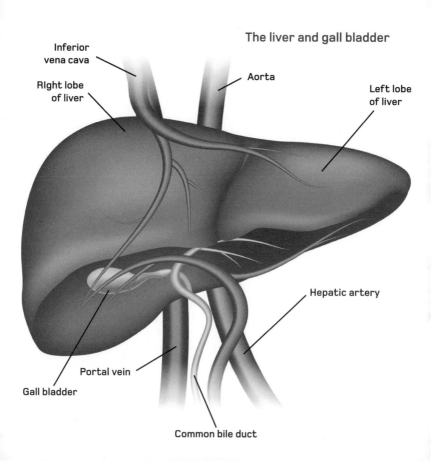

The liver and gall bladder

Inferior vena cava

Aorta

Right lobe of liver

Left lobe of liver

Hepatic artery

Portal vein

Gall bladder

Common bile duct

The large intestine

From the ileum, the watery remains of food enter the large intestine (the colon). This is about 1.5 metres (4.9 ft) long and generally twice as wide as the small intestine. The ileum connects at the base of the colon, where there is a lower pouch-like section, called the caecum, which is a vestige of the large pouches used by herbivores to store food as it is fermented by bacteria. In humans they are much reduced in size, but still act as stores of useful gut bacteria that help digest foods that have made it through the small intestine.

The chyme (digested material) is pushed up and around the colon, which forms a loop around the outside of the small intestine. The colon has four sections: ascending, transverse, descending and the sigmoid colon, which connects to the rectum. It takes about 16 hours for the chyme to travel through the colon. In this time, the water and vitamins used in digestion are reabsorbed, along with any nutrients released by gut bacteria. As it reaches the rectum, the chyme becomes more solid.

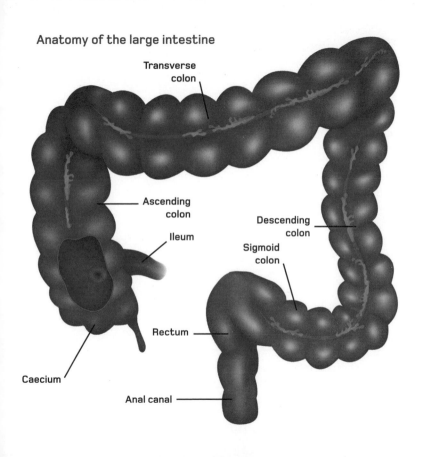

Anatomy of the large intestine

Transverse colon

Ascending colon

Ileum

Descending colon

Sigmoid colon

Rectum

Caecium

Anal canal

Peristalsis

Being a convoluted tube, the digestive system cannot rely on gravity to pull food through it. Instead it uses a form of rhythmic muscle contractions called peristalsis. This type of involuntary motion takes place in other parts of the body, too – for example, during the ejaculation of sperm.

In the digestive system, peristalsis begins with swallowing and continues by pushing packets of food (boli; sing. bolus) down the oesophagus, to the stomach – during vomiting it pushes the other way. Peristalsis is also intrinsic to the functioning of the intestines, since it mixes the chyme while pushing it along. Here the chyme is more watery and more difficult to push, which is why fibre is important in the diet, to give the intestines something to act against. Peristalsis is performed by circular smooth muscles that ring the alimentary canal. These contract and relax in a rhythmic fashion, creating narrowings of the canal that move along in waves, pushing boli with them. Should a bolus get stuck, stretch receptors stimulate stronger contractions in that area.

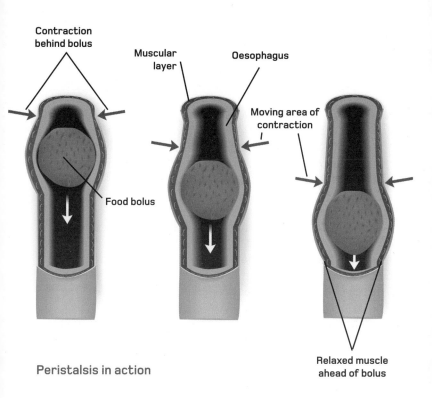

Contraction behind bolus

Muscular layer

Oesophagus

Moving area of contraction

Food bolus

Relaxed muscle ahead of bolus

Peristalsis in action

Defecation

The alimentary canal ends at the rectum. This chamber is about 12 centimetres (4¾ in) long, and a vertical extension of the sigmoid colon – 'rectum' means 'straight'. It widens at the base, forming the rectal ampulla, which is where faeces, the solid waste left over from digestion, is temporarily stored.

As faeces builds up, the stretching of the rectal wall creates an urge to expel it. If this is not acted upon, the faeces is pushed back into the colon, where it is desiccated further, making it harder, and ultimately more difficult to expel. To exit the body the faeces must pass through the anal canal. This 3–4 centimetre (1–1½ in) section has a double sphincter; the upper is involuntary, the lower one voluntary. As the faeces enters the canal, peristalsis in the rectum pushes it through. The faeces has a coating of mucus to help it slide better. The doubling up of sphincters ensures that the rectum is sealed off from the faeces before it leaves the body. Gases produced by gut bacteria leave the same way, as wind.

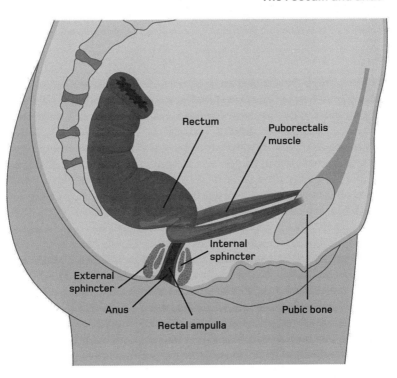

Rectum

Puborectalis muscle

Internal sphincter

External sphincter

Anus

Rectal ampulla

Pubic bone

The respiratory system

The human body requires a more or less constant supply of oxygen. That oxygen is sent to every cell in the body, where it is used to release energy from sugars. The waste products are water and carbon dioxide. The role of the respiratory system is to collect oxygen from the air, and expel the waste carbon dioxide. The body can extract energy without using fresh oxygen, for a short time at least, in a process called anaerobic respiration. The waste product of this is lactic acid, which literally burns the muscles as it builds up. Oxygen is needed to clear away this ultimately toxic material.

The respiratory system is centred on the lungs, in the chest. Airways connecting to the mouth and nose allow gases to move in and out of the lungs. Fresh air coming into the lungs gives up some of its oxygen, and takes carbon dioxide from the body in a process called gas exchange. The stale air then leaves the body. Humans use an active form of breathing to pull and push air in and out of the lungs.

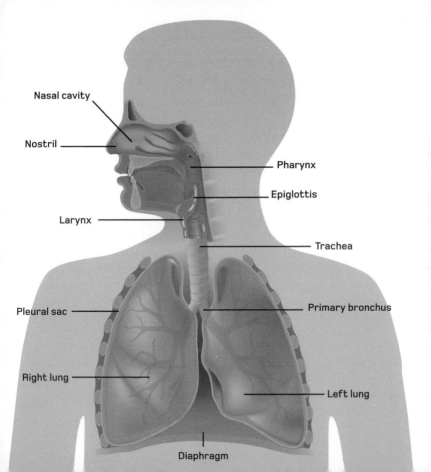

The lungs

Air arrives in the lungs via the trachea or windpipe, having entered through the mouth or nostrils. The trachea starts at the pharynx, where it diverts from the oesophagus. (During swallowing, a cartilage flap called the epiglottis seals the trachea to prevent food choking off the air.) Just before reaching the lungs, the airway splits into two bronchi.

Each bronchus delivers air to one lung. Once inside the lung, the bronchus divides hundreds of times to create a root-like structure containing smaller airways called bronchioli. These are lined with mucus, which traps dirt in the air. Beneath that is a layer of cells with hair-like cilia that waft used mucus out of the lung so it can be coughed up and swallowed. The bronchioli also have a layer of muscle that dilates or contracts the airway, to control how much air gets through. The smallest bronchioli are 0.5 millimetres wide and terminate in pouches called alveoli. It is here that gas exchange occurs. Each lung contains two million alveoli, giving it a spongy structure.

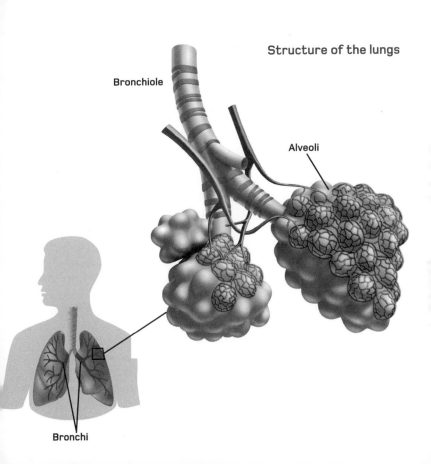

Structure of the lungs

Bronchiole

Alveoli

Bronchi

Gas exchange

The process that brings oxygen into the blood, for dissemination through the body, and removes carbon dioxide waste is called gas exchange. It takes place in the alveoli – tiny pouches that fill the lungs and are surrounded by capillaries. These tiny blood vessels bring deoxygenated blood – that is, blood high in carbon dioxide – very close to the lining of the alveolus, which is one cell thick.

Fresh air from an in breath enters the alveolus. Some of the oxygen dissolves in the fluid lining the pouch. The concentration of oxygen in the air is much higher than in the blood, so oxygen molecules diffuse through the alveolal lining and enter the blood plasma. The oxygens are bonded to haemoglobins, iron-rich proteins that give red blood cells their colour. The blood has now become oxygenated. Carbon dioxide is dissolved in the blood plasma at concentrations much higher than the air in the alveolus. As a result, it diffuses out into the air pouch, where it is exhaled as part of the remaining air.

Inside the alveolus

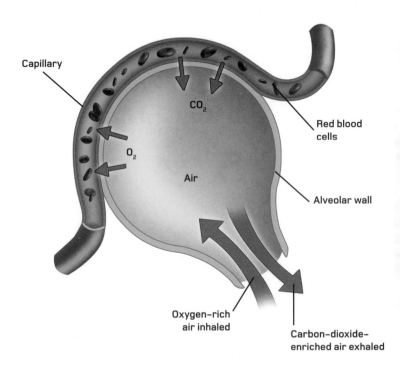

Capillary

CO₂

O₂

Air

Red blood cells

Alveolar wall

Oxygen-rich air inhaled

Carbon-dioxide-enriched air exhaled

Breathing

The average human breathes 15 times a minute, taking in about 13,500 litres of air every day. As it comes in, the air is 21 per cent oxygen with only a tiny amount of carbon dioxide present. Following gas exchange, exhaled air is about 5 per cent carbon dioxide and 15 per cent oxygen.

Human breathing is reciprocating, which means that air must flow in, stop, then flow out again. This cycle is achieved using a domed muscle called the diaphragm, which hangs from the spine and ribs to divide the internal body cavity into two. To inhale, the diaphragm contracts and flattens. This increases the volume of the thoracic cavity above, reducing the pressure inside. The air outside is now at a higher pressure and so rushes in, filling the lungs. To exhale, the diaphragm relaxes, and returns to its domed shape. That shrinks the thoracic cavity, increasing its internal pressure, and the air in the lungs, now depleted of oxygen, is pushed out. To take deep breaths, the intercostal muscles in the ribs enlarge the thoracic cavity further by pulling the ribcage up and out.

Inhalation and exhalation

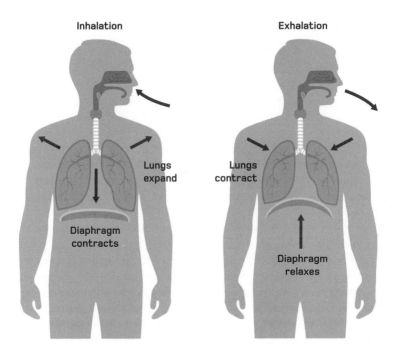

Coughing

If an unwanted object enters the trachea, a coughing fit will ensue to clear it. The cough could be stimulated by a food item being lodged in the windpipe, some saliva trickling in by chance or, if you have a cold, thick phlegm dripping down from the nasal passages. Although we can clear the throat with a voluntary cough, the most powerful and effective coughs are involuntary reflexes. They propel air out of the lungs and up the windpipe at 160 kilometres an hour (100 mph), which is generally enough to clear the blockage.

The coughing reflex begins with the pharynx and vocal cords opening wide to accommodate a large and sudden inhalation that fills the lungs. The epiglottis closes (in a normal breath it remains open), while the diaphragm relaxes and the abdominal muscles contract to push the diaphragm higher into the thoracic cavity, ramping up the pressure. The build-up forces the epiglottis to open, releasing a rush of air that clears the blockage. As it blasts past the vocal cords, the air creates the loud telltale sound.

The cough reflex

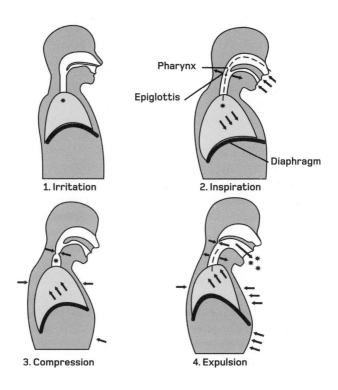

1. Irritation

2. Inspiration
- Pharynx
- Epiglottis
- Diaphragm

3. Compression

4. Expulsion

Sneezing

A sneeze is a reflex action that clears a blockage in nasal passages. It is stimulated by irritation in the lining of the nose, indicating there is something that needs to be removed.

A sneeze begins with a series of short in breaths that fill the lungs to maximum capacity. The epiglottis closes, sealing off the airway, and the diaphragm and rib muscles spasm, causing the pressure to rise. When a high pressure is reached, the epiglottis opens and the air leaves the windpipe at around 160 kilometres an hour (100 mph), as in the cough reflex (see page 96). However, the air has not reached its target yet, and the tongue and soft palate partially seal the back of the mouth, so that most of the air is directed through the nostrils – sending out an aerosol of mucus at the same time. The speed of the air from the nose has slowed, but is still around 70 kilometres an hour (45 mph). The sneeze reflex tends to contract many upper body muscles, including the eyelids, although it is still possible to sneeze with your eyes open, contrary to popular belief.

Sneezing in action

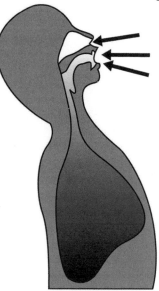

Nerves inside the nostril sense an inhaled irritant and the brain signals to the lungs to inhale deeply.

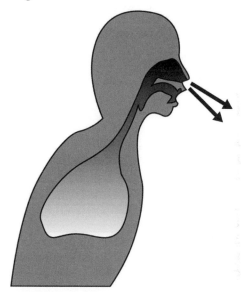

The airways close and air pressure builds in the lungs. A sudden release of pressure results in a sneeze.

Hiccups

Breathing in and out – and sneezing, coughing and laughing – require a series of precisely sequenced steps. When these steps get out of sync, you end up with the hiccups. As the diaphragm contracts and flattens, it draws air into the windpipe and lungs. At this point the epiglottis and vocal cords are meant to be wide open to allow the air to rush in. However, if they close during inhalation, the last section of air to pass through makes the vocal cord produce a hiccup. In this respect a hiccup is a reverse cough; in a cough, the epiglottis and vocal cords shut as the exhalation process begins.

Why we hiccup is something of a mystery. It appears to be linked to having a full stomach, and so the hiccup reflex might be the body's way of creating a sudden drop in pressure that pulls air bubbles trapped in the stomach up to the oesophagus – much like a burp. Another suggestion is that hiccups are part of the fetal development of nerve control of the diaphragm that persists into adulthood.

Hiccups in action

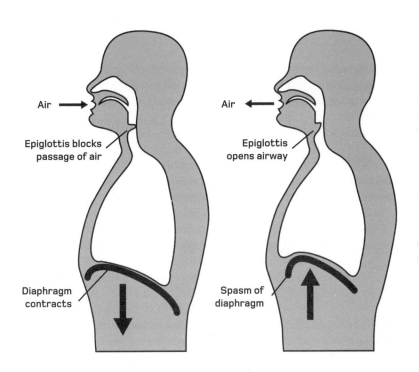

Air

Epiglottis blocks
passage of air

Diaphragm
contracts

Air

Epiglottis
opens airway

Spasm of
diaphragm

The cardiovascular system

Containing the heart and blood vessels, the cardiovascular system is the body's primary transport system. It holds approximately 5 litres (8 pt) of blood, which is a mixture of liquids and cells. The heart pumps all this blood around the body, making a complete circuit in less than one minute.

The system is composed of three types of blood vessel: arteries, veins and capillaries. All are hollow tubes, but are structured differently according to the direction and content of the blood they carry. The blood has multiple roles. It contains elements of the immune system, ready to tackle infections. The vessels also link to the body's main glands, and carry the hormones and other chemical signals that they secrete. Nutrients from the intestines enter the body via the blood, and travel to the liver along a dedicated vein. From there, the blood circulates the nutrients to the rest of the body. The most crucial – at least, the most constant – job of the cardiovascular system is to transport oxygen collected by the lungs to every cell in the body.

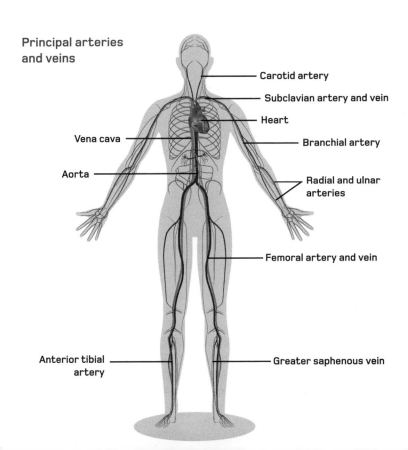

Principal arteries and veins

Carotid artery

Subclavian artery and vein

Heart

Branchial artery

Vena cava

Aorta

Radial and ulnar arteries

Femoral artery and vein

Anterior tibial artery

Greater saphenous vein

Blood

Human blood is a mixture of water and cells, plus several hundred chemicals in solution or suspension. The constituents of blood divide into three sections: plasma, the buffy coat and erythrocytes, or red blood cells. These sections become apparent when blood is separated in a centrifuge, resulting in the heaviest items (blood cells) moving to the bottom of the sample and the lightest (plasma) floating to the top.

The red blood cells comprise 45 per cent of the total volume, and it is these that give blood its red appearance. The buffy coat is a pale band of material making up less than one per cent of the total volume. It consists of white blood cells and platelets, which are involved in blood clotting. The remaining 55 per cent is plasma. This yellow liquid is 92 per cent water, with the remaining eight per cent largely composed of proteins, most significantly serum albumins. These thicken the blood and increase its surface tension, thus ensuring that the blood remains a coherent substance as it circulates.

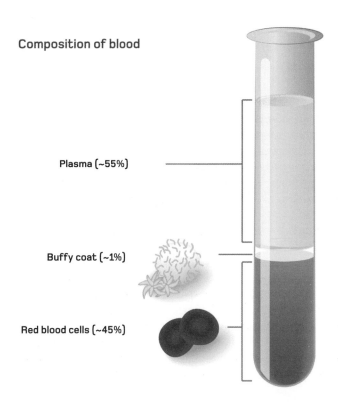

Composition of blood

Plasma (~55%)

Buffy coat (~1%)

Red blood cells (~45%)

Red blood cells

Also known as erythrocytes ('red cells' in Latin), there are five billion red blood cells in every millilitre of blood. Their role is to carry oxygen from the lungs to the rest of the body. They achieve this using chemicals called haemoglobins. The *-globin* term refers to globular proteins that make up most of this substance, while *haemo-* means 'iron'. The red colour comes from the presence of this metal in the molecules.

Red blood cells are entirely devoted to carrying oxygen. There is no room for a cell nucleus and 96 per cent of the dry weight of the cell is from haemoglobin. Each haemoglobin can carry four oxygen molecules at a time. When it gives away its oxygen cargo, the haemoglobin picks up some carbon dioxide to transport back to the lungs. However, most of this waste material simply dissolves in the blood plasma. Red blood cells are disc shaped with a dented surface. This maximizes their surface area for collecting and releasing oxygen. On average a red blood cell will be in circulation for 100 days before it is removed by the liver.

Red blood cells make up
70 per cent of the total
body cell count.

Bone marrow

All blood cells, red and white, are produced in the marrow inside large bones. Red blood cells carry oxygen; the many kinds of white cell work for the immune system; and tiny cells called platelets are used for blood clotting.

The process of producing blood cells is called haematopoiesis. It occurs mostly in the large leg bones during childhood, but from the age of about 25, the marrow in the pelvis, sternum, vertebra and ribs become the main sites. Red blood cell production is stimulated by the hormone erythropoietin. This is produced by the kidneys when they detect a fall in the amount of oxygen in the blood – an indication that more red blood cells are needed. The red blood cells form from a haemoblast, a type of stem cell (see page 256) that can produce any kind of blood cell. The stem cell divides several times, and after each division it becomes more specialized. During the proerythroblast stage it begins to accrue haemoglobin. In the next stage, the erythroblast, the nucleus is expelled, and it finally divides into erythrocytes.

Formation of blood cells

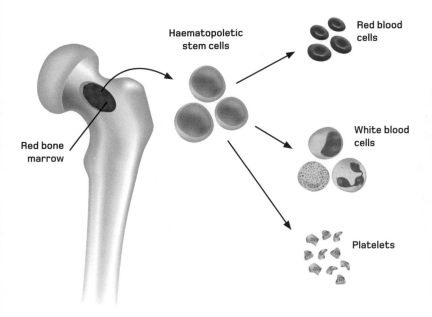

Haematopoietic stem cells

Red bone marrow

Red blood cells

White blood cells

Platelets

The heart

The human heart is a fist-shaped organ in the middle of the chest. It weighs about 300 grams (10½ oz) and is made up of four hollow chambers, each surrounded by a thick muscular wall. The heart's tissue is largely cardiac muscle, which is able to contract strongly without becoming fatigued in the same way that skeletal muscle does. The two upper chambers are atria, while the two lower, and larger ones, are ventricles. Blood enters the heart through the atria and leaves via the ventricles.

The heart is generally considered in two halves, divided between left and right. The left side receives oxygenated blood from the lungs and pumps it out to the rest of the body. The right side receives deoxygenated blood from the body and sends it to the lungs for an oxygen refill. The left chambers are larger and more muscular to account for their bigger job. The entry and exit points of both ventricles are sealed with valves. These close off to allow the beating heart to pressurize the blood adequately enough to force it through the vessels.

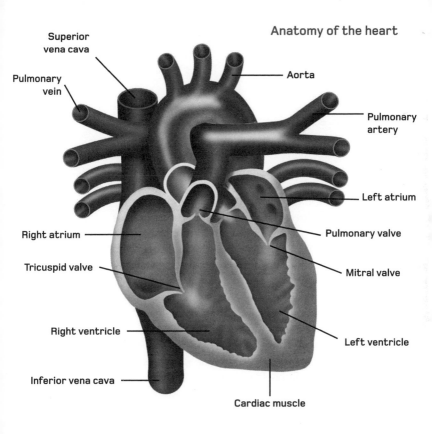

Anatomy of the heart

Superior vena cava

Pulmonary vein

Aorta

Pulmonary artery

Left atrium

Pulmonary valve

Right atrium

Tricuspid valve

Mitral valve

Right ventricle

Left ventricle

Inferior vena cava

Cardiac muscle

The heartbeat

The heartbeat is often characterized as a double-tone, or 'lub-dub' sound. This is because a wave of contraction runs through the heart from top to bottom, with the atria contracting just before the ventricles. The beat is actually a five-stage cardiac cycle of contractions and relaxations.

The cycle is initiated by the sinoatrial node, or cardiac pacemaker, on the right atrium. This produces an electrical pulse that travels through the heart muscle. The frequency of the cardiac cycle (heart rate) is influenced by nervous impulses at the pacemaker, sent by the brain and spinal cord. In the first stage of the cycle the heart is relaxed, and the valves between the atria and ventricles open. Then the atria contract, pushing blood through into the ventricles. The atrioventricular valves close, and the ventricles begin to squeeze the blood. In stage four, the semilunar valves at the exits to the ventricles open, pushing the blood up and out. In the final step, the semilunar valves close to stop blood seeping back in, and the empty ventricles begin to relax.

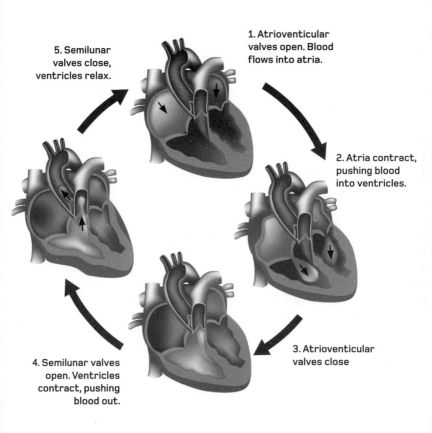

5. Semilunar valves close, ventricles relax.

1. Atrioventicular valves open. Blood flows into atria.

2. Atria contract, pushing blood into ventricles.

3. Atrioventricular valves close

4. Semilunar valves open. Ventricles contract, pushing blood out.

The circulatory system

The cardiovascular system has a double circulatory structure. Simply put, it involves two loops of vessels, both emanating from the heart. Each loop is made from three kinds of vessel: arteries carry blood away from the heart, while veins bring it back. In between, smaller capillaries permeate the tissues.

The systemic circulation is the larger loop, delivering oxygen to the body. It begins at the aorta, the primary artery, with the largest vessel being about 2 centimetres (¾ in) wide. The aorta emerges from the left ventricle and loops round into the abdomen, with various arteries branching off. The systemic system returns blood to the heart at the left atrium via two massive veins – the superior and inferior vena cavae. The second loop, pulmonary circulation, is shorter and connects the heart to the lungs. The pulmonary artery carries deoxygenated blood delivered by the vena cava from the right ventricle to the lungs. The pulmonary vein then returns oxygen-replenished blood to the left atrium, where it enters the systemic system.

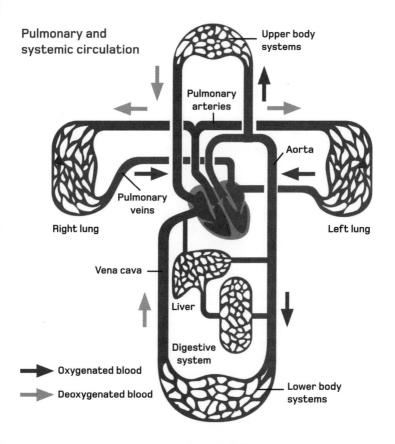

Pulmonary and systemic circulation

Upper body systems

Pulmonary arteries

Aorta

Pulmonary veins

Right lung

Left lung

Vena cava

Liver

Digestive system

Lower body systems

→ Oxygenated blood

→ Deoxygenated blood

Arteries

The arteries carry blood away from the heart, which means they transport oxygenated blood – with one exception: the pulmonary artery carries deoxygenated blood.

Arteries tend to be located deep in the body to protect them and their oxygen supply from injury. Veins are nearer the surface and more easily visible. High levels of oxygen in arterial blood give it a brighter red colour than venous blood. An injury that releases bright red blood has therefore caused damage to an artery and is more serious. The cavity, or lumen, of an artery tends to be smaller than that of a vein. This is due to a thicker layer of smooth muscle and an elastic layer that makes arteries the firmer kind of vessel. Arterial muscles contract and dilate the lumen to help regulate blood pressure in the vessel. The blood pressure inside arteries rises and falls in time with the beating of the heart. The surge of pressure that comes with each beat can be felt as a pulse at various points around the body, where arteries are close to the skin.

Structure of an artery

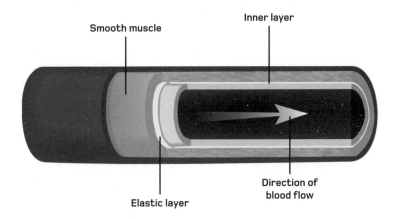

Smooth muscle

Inner layer

Elastic layer

Direction of
blood flow

Veins

With the exception of the pulmonary vein, which connects the lungs to the heart, veins carry deoxygenated blood away from body tissue and back to the heart. A vein has three main layers, with the innermost including one-way flap valves that allow blood to move in just one direction. A middle muscle layer expands and contracts the vessel to control body heat, while the outer layer is mostly connective tissue.

Superficial veins carry blood just below the skin; when the body is overheated, the veins dilate, enriching the skin with blood and releasing heat. To conserve heat, the veins contract, making the skin pale. Although superficial veins appear blue through the skin, the blood within them is actually dark red. This darker colour is due to the blood's haemoglobin bonding to carbon dioxide, rather than oxygen. However, 75 per cent of the carbon dioxide content is actually dissolved in the blood plasma rather than carried by the haemoglobin, and this makes venous blood more acidic than arterial blood.

Structure of a vein

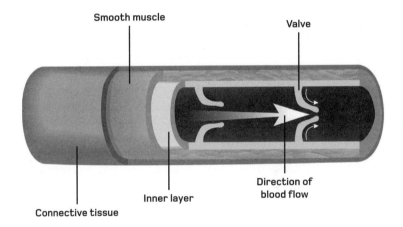

Capillaries

Connections between arteries and veins are made by capillaries. These blood vessels are very fine – just a few thousandths of a millimetre wide. They branch out from small arteries, or arterioles, to form an intricate, interweaving network of vessels, known as a capillary bed. The tiny vessels then converge again to connect to a venule, a small vein that feeds blood into the venous system.

Capillary beds permeate the organs and tissues to provide oxygen, nutrients and hormonal signals to every cell in the body. The more energetic structures, such as the muscles and digestive system, are served by more extensive beds. A capillary has no sheathing with muscle or connective tissue. In fact, the capillary wall is one cell thick. Small molecules – water and important gases – move in and out of capillaries through the gaps between cells. The chemistry of the blood varies through the bed to encourage fluid, and the gases it contains, to diffuse out to release oxygen and to diffuse in to collect waste carbon dioxide.

Capillary network

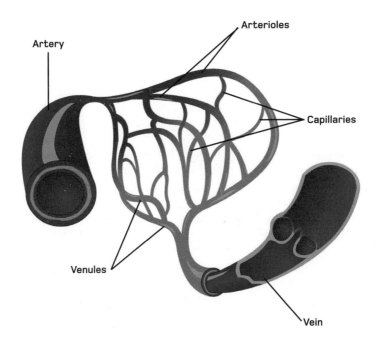

Artery

Arterioles

Capillaries

Venules

Vein

The immune system

The human body is under constant attack, and is defended by the immune system, a somewhat nebulous collection of tissues that all have a role to play. It is geared to prevent the entry of pathogens (disease-forming agents) and other foreign bodies, to identify any that do get past the defences, and then to destroy invaders without damaging healthy body tissue.

The skin and epithelial tissues – the lining of the airways and other tracts entering the body – form a physical barrier to invaders. For example, nose hairs filter out dust from the air, while sticky mucus lining the throat and elsewhere traps smaller particles and is systematically removed and replaced with a clean supply. Under severe attack, these surfaces become inflamed, boosting mucus production and bringing more blood to the scene. The blood contains white blood cells, which work to block any breaks in the body's defensive barrier. They also learn to recognize an invader, so that it can be hunted down and destroyed throughout the body.

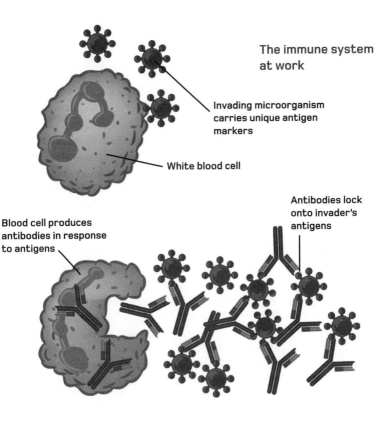

The immune system at work

Invading microorganism carries unique antigen markers

White blood cell

Antibodies lock onto invader's antigens

Blood cell produces antibodies in response to antigens

Clotting

If a blood vessel in the skin is damaged, it creates an opening through which infection can get into the body. In order to prevent this, the blood clots, or coagulates, to create a temporary seal until the skin can repair itself.

Clotting is controlled, in part, by tiny blood cells called platelets, or thrombocytes. These mass into a damaged area to form an initial plug. They become activated, changing from a globular shape to a spiky stellate one to cling together. The damaged tissue around the broken blood vessel, and other white blood cells on the scene, release a chemical called tissue factor, triggering a complex cascade of chemical changes in the blood. In simple terms, a blood chemical called prothrombin breaks apart to form the active hormone thrombin. This acts on a soluble protein present in the blood called fibrinogen, making it transform into insoluble fibrin. The addition of fibrin steadily thickens the blood into a gel. Finally a solid lattice of fibrin strands builds up around the platelet mass forming a scab, sealing the damage completely.

The clotting process

Skin damage

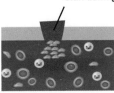

1. Platelets accumulate at wound.

2. Platelets break down to release proteins.

3. Proteins break down prothrombin into thrombin.

4. Thrombin acts on fibrinogen to produce blood-thickening fibrin.

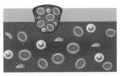

5. Fibrin creates scab around platelets.

White blood cells

The white blood cells are the patrol force of the body. They are made in the bone marrow from the same kinds of stem cell that produce red blood cells and platelets. Unlike the other blood cells, all white blood cells, or leucocytes, have nuclei.

There are five broad categories of leucocyte. Neutrophils are the most common, making up about 60 per cent of the total. These cells are highly mobile and wriggle out of capillaries to get at bacteria in surrounding tissues. The neutrophils are phagocytes – in other words, they engulf and digest attackers. The primary role of eosinophils (about 3 per cent) is releasing chemical weapons to attack larger infectious agents, like worms. Basophils, the rarest white blood cells (less than 1 per cent), initiate inflammation to combat infections. Lymphoctyes (about 30 per cent) are the immune system's detectives. They seek out and identify infectious agents and cancers, creating antibody tags for them. The final kind of leucocyte, monocytes (about 7 per cent), are the destroyers sent to engulf anything tagged for removal.

Types of white blood cell

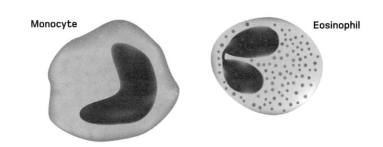

Monocyte

Eosinophil

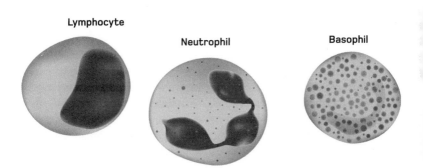

Lymphocyte

Neutrophil

Basophil

Antibodies

All but the most dangerous pathogens (see pages 368 and 370) betray themselves. All cells have a series of identifying marks on their surface – chemical flags called antigens. Cells in the human body use a common set of markers, known as the human leukocyte antigens. Invaders do not have these markers, although a few, such as the AIDS virus and liver flukes, have ways of fooling the system.

When a lymphocyte encounters a pathogen, it creates an antibody to match the invader's foreign antigens. The antibody is a large protein that is shaped to lock onto the antigens. Once a lymphocyte has created an antibody, it divides into plasma cells, which mass-produce it. The antibody is released into the blood, where it attaches to anything that shows the matching antigen. Antibodies can then do a number of things: they may simply stop the pathogen causing disease; others glue pathogens together for engulfment by monocytes; and sometimes the antibody calls in other chemicals to disintegrate the threat.

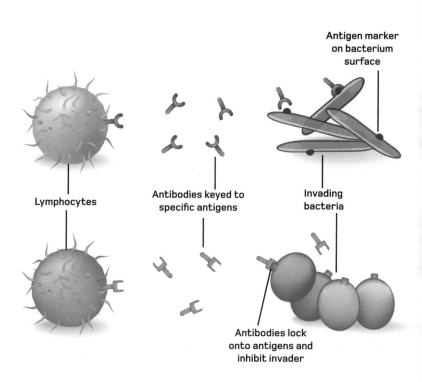

Lymphocytes

Antibodies keyed to specific antigens

Antigen marker on bacterium surface

Invading bacteria

Antibodies lock onto antigens and inhibit invader

Immunity

There are some illnesses that you can only catch once – you become immune to further infections. The familiar examples are chickenpox, measles, mumps and rubella. If you are exposed to any of these viruses a second time, the immune system deploys the antibodies it used the first time (see page 128) and the infection never becomes pathological. In fact, you can only catch a common cold or flu once as well. The problem is that the viruses that cause them are constantly mutating into new forms that the immune system must learn to defeat each time.

A lymphocyte, the cell type that makes antibodies during an infection, generally circulates in the blood for no more than a few days. (A short, but active, life is true for most kinds of white blood cell.) However, once an antibody is identified and put into mass production by plasma cells, the parent lymphocyte also creates a small number of memory cells. These live in the blood stream for years, holding the information required to fight off the same infection should it reoccur.

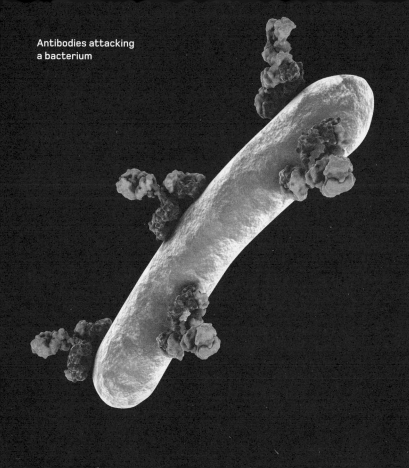
Antibodies attacking
a bacterium

Vaccinations

A vaccination, or inoculation, is a procedure that stimulates immunity to a disease, without a person having to suffer the infection in advance. Vaccinations are used for dangerous diseases, such as typhoid and rabies, and widespread ones, such as measles, which are mostly harmless but do carry a risk of life-changing complications.

The term 'vaccination' derives from the Latin word for cow. In 1796, Edward Jenner, an English doctor, noticed that dairy workers previously infected with the mild disease cowpox never caught the deadlier disease smallpox, which claimed many lives in those days. He injected a boy with pus from a cowpox lesion. The patient had a mild illness for a few days. Two months later, Jenner injected the same boy with smallpox – and the child was unaffected. Jenner's technique was to stimulate the immune system to make antibodies for smallpox using a weakened version of the disease. Modern vaccines work the same way, either using a 'live' vaccine or a chemical substitute of the relevant antigen.

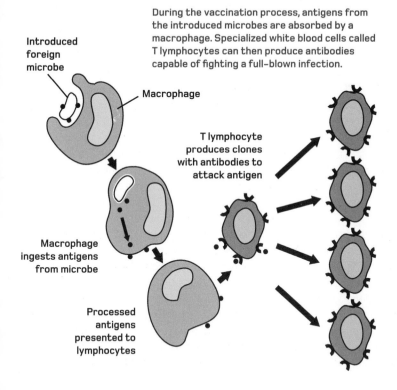

During the vaccination process, antigens from the introduced microbes are absorbed by a macrophage. Specialized white blood cells called T lymphocytes can then produce antibodies capable of fighting a full-blown infection.

Introduced foreign microbe

Macrophage

T lymphocyte produces clones with antibodies to attack antigen

Macrophage ingests antigens from microbe

Processed antigens presented to lymphocytes

Inflammation

When a part of the body is compromised, it becomes inflamed. This is the immune system's generic response to attacks of all kinds, including infections, burns, cuts and bruising.

Inflammation has five primary characteristics: the affected region swells as blood vessels dilate to let in a greater volume of blood. The region also becomes red and warm, and some or all of the function of the body part is lost to help with healing. Finally, it hurts. The inflammation response is controlled by many factors, one of which is histamine. This makes the nerve endings serving the inflamed region more sensitive to avoid further injury. It can also make the region itch, the scratching of which is thought to initiate immune activity. In epithelial tissue – the lining of the nose, for example – histamine stimulates extra mucus production to wash away pathogens. Temperature increase, a clear sign of infection, allows the immune system to work faster to tackle a threat, while swelling closes cuts and brings in more white blood cells to seek and destroy invaders.

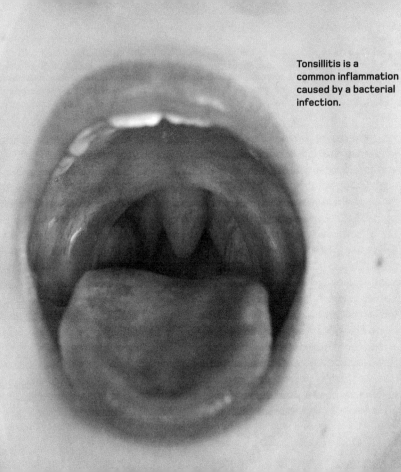

Tonsillitis is a common inflammation caused by a bacterial infection.

Allergies

An allergy is the immune system being overly sensitive to a substance, known as an allergen, entering the body. The immune system identifies the harmless or mostly harmless allergen as being a danger and launches an inappropriate level of inflammation. Common allergens are pollen and other airborne particles; foods, often nuts or shellfish; and insect stings.

Antibodies aimed at these allergens also lock on to basophil cells, triggering the release of histamines. The results can range from an itchy skin rash or a runny nose, to dangerous swelling around the airways and in the lungs. In extreme cases, a plunge in blood pressure results in a heart attack. Such life-threatening allergic reactions are known as anaphylaxis. Allergies can limit a sufferer's ability to live normally and the causes are poorly understood. Treatments involve tackling inflammation with antihistamines, which block histamine production, and steroids. Anaphylaxis requires the administration of adrenaline to raise blood pressure.

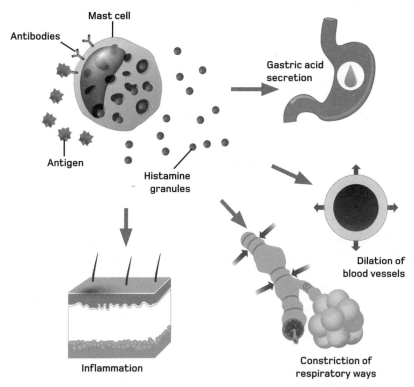

Allergic responses are triggered when a specialized white blood cell called a mast cell releases histamines in response to an antigen.

Mast cell

Antibodies

Antigen

Histamine granules

Gastric acid secretion

Dilation of blood vessels

Inflammation

Constriction of respiratory ways

The lymphatic system

The lymphatic system is a network of vessels interwoven with blood supply. One of its roles is to collect plasma that leaks out from capillaries. Most of that liquid finds its way back into blood vessels, but the remainder stays in the tissues. This fluid drains away through the lymphatic system as a colourless liquid called lymph. There is no pumping mechanism to push lymph. Instead, it is squeezed along by the general motion of the body. Valves prevent the liquid sinking to the feet, and the lymph slowly makes its way into a large duct along the spine before feeding back into the blood through veins near the neck.

The other function of the system is immunological. Lymph nodes are filters located on major lymphatic vessels. They contain white blood cells that remove any pathogens in the passing lymph. Large lymph nodes are located in the armpits, groin and neck. During an infection, the number of white blood cells increases significantly, packing the nodes tight – a painful symptom of illness described as 'swollen glands'.

Anatomy of the lymph system

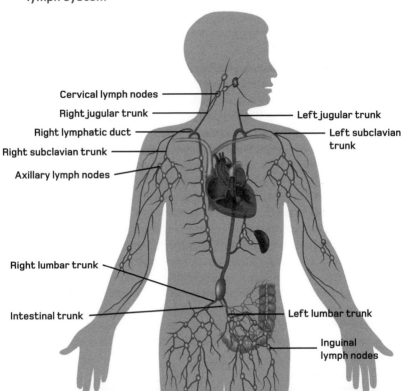

Cervical lymph nodes

Right jugular trunk

Right lymphatic duct

Right subclavian trunk

Axillary lymph nodes

Left jugular trunk

Left subclavian trunk

Right lumbar trunk

Intestinal trunk

Left lumbar trunk

Inguinal lymph nodes

The nervous system

The nervous system is the body's chief coordination controller. Its most obvious function is in the collection and perception of sensory information and the voluntary responses to it – in the form of muscle, or motor, control. However, a hidden aspect to the nervous system also controls the involuntary muscles and influences the actions of glands and organs.

The nervous system is primarily made up of specialized cells called neurons. These carry signals in the form of electric pulses. The system as a whole is best understood when divided into a series of subsystems. The central nervous system is made up of the brain and spinal cord. This is where sensory information is received and from where any responses are sent. The peripheral nervous system is a network of nerves throughout the body. It collects signals from sense organs and transmits any responses. The peripheral system is further divided into the somatic (voluntary) system and the autonomic (involuntary) system.

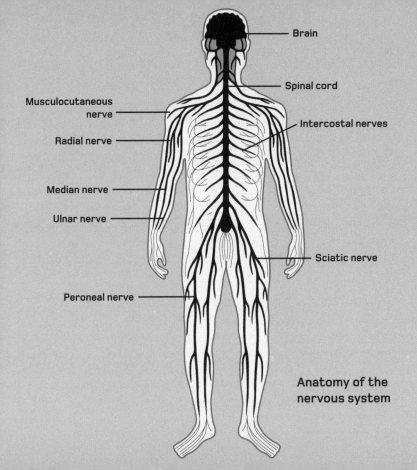

Brain

Spinal cord

Musculocutaneous nerve

Intercostal nerves

Radial nerve

Median nerve

Ulnar nerve

Sciatic nerve

Peroneal nerve

Anatomy of the nervous system

Neurons

The nervous system is built up of nerve cells, or neurons. These cells are present in huge quantities in the brain and spinal cord. Elongated versions form the nerves that make up the peripheral system, and a wide variety of modified neurons are used in the sense organs.

The typical neuron has a cell body, in which the cytoplasm, nucleus and other machinery are present. The cell membrane also extends into multiple branches. The longest and thickest branch is called the axon, while the multiple smaller branches are dendrites. The axon is akin to a conducting wire and carries the cell's electrical impulses. In long nerves, the axon is sheathed in a fat called myelin, which insulates the signal and speeds it up. At the far end, the axon splits into several terminals. These do not physically connect to the next neuron. Instead, nerve signals are sent across a tiny gap, or synapse, as chemicals called neurotransmitters. These enter a dendrite of a neighbouring neuron, and the signal continues its journey through that cell as an electric pulse.

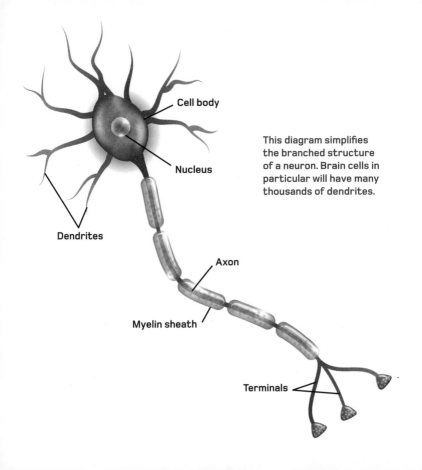

Cell body

Nucleus

Dendrites

Axon

Myelin sheath

Terminals

This diagram simplifies the branched structure of a neuron. Brain cells in particular will have many thousands of dendrites.

Action potential

A nerve signal is not an electric current nor a spark that flashes through a neuron. It is an electrical potential, or a difference in charge between the inside and outside of the cell.

When resting, an axon has a negative charge inside and is positive on the outside. Pumps in the membrane push out sodium ions (Na^+), while chloride ions (Cl^-) are kept inside. The same pumps push potassium ions (K^+) in, but at a slower rate, thus creating the charge difference. A chemical stimulus from the cell body opens channels in the axon that allow Na^+ to flood inside, flipping the charges. The sodium channels then close and the potassium ones open. The K^+ ions now flow out, flipping the charges all over again. The system then reverts to normal, with ions being pumped in and out to restore the resting state. However, the flip charge has an effect on the next section of axon, causing its sodium channels to open and for the process to repeat again and again along the axon. This creates a so-called action potential that sweeps through the axon.

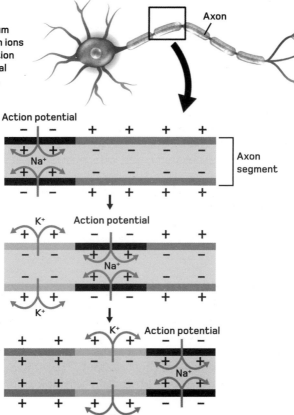

Flows of sodium and potassium ions sweep the action potential signal along an axon.

The somatic nervous system

This part of the peripheral nervous system forms the link between our senses and the skeletal muscles that control posture and movement. The system is made up of two sets of complimentary nerves. Afferent nerves bring information from the body to the central nervous system, while efferent nerves carry signals out from the brain and spinal cord to the body. Afferent nerves are sensory – they relay information picked up from sense organs. Efferent nerves are motor nerves, in that they innervate skeletal muscles, which contract in response to what the senses have detected.

For every afferent sensory nerve there is a corresponding efferent motor one, and these paired nerves are organized into 43 sets: 12 are cranial nerves, which connect directly to the brain. They are associated with the sense organs on the head and facial and mouth muscles. The other 31 sets are spinal nerves. These collect sensory inputs from the skin, and control the voluntary muscles from the neck down.

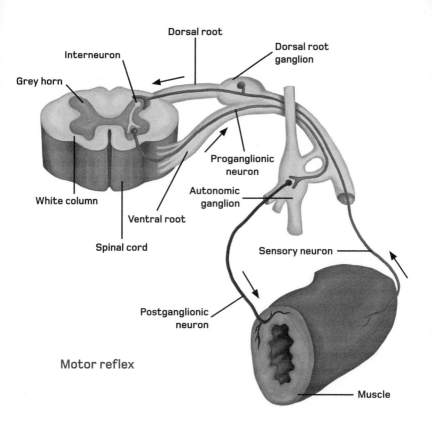

Dorsal root

Dorsal root ganglion

Interneuron

Grey horn

Proganglionic neuron

White column

Autonomic ganglion

Ventral root

Sensory neuron

Spinal cord

Postganglionic neuron

Motor reflex

Muscle

The spinal cord

The spinal cord forms the lower section of the central nervous system. It extends from the medulla oblongata in the brainstem, down inside the vertebral column, and ends between the first and second lumbar vertebra in the lower back.

About 44 centimetres (17 in) long on average, the spinal cord is a bundle of nervous tissue. It forms a high-speed pathway between the peripheral nervous system and the brain, but it is also involved in reflex actions – closed loops of stimulus and response that do not require voluntary input from the brain. The outer region of the cord is white matter, dominated by axons (see page 142), and is used for carrying signals. The interior, where nerve cells are more compact, is grey matter, which processes signals coming in from the white matter. The nerves of the peripheral nervous system connect on both sides of the spinal cord, passing through the white matter to the grey matter. The afferent sensory nerves bringing signals are connected behind the efferent motor nerves, which carry signals out to the body.

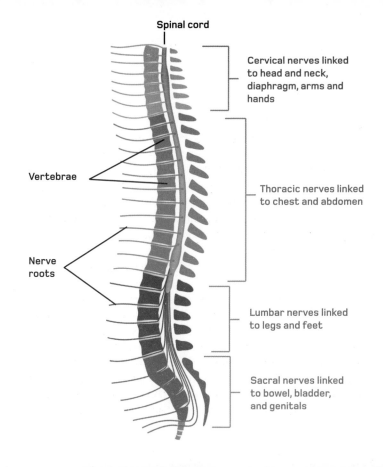

Spinal cord

Cervical nerves linked to head and neck, diaphragm, arms and hands

Vertebrae

Thoracic nerves linked to chest and abdomen

Nerve roots

Lumbar nerves linked to legs and feet

Sacral nerves linked to bowel, bladder, and genitals

Reflexes

A reflex is an automatic muscle response to a specific stimulus. It involves skeletal muscles that are normally under voluntary control, but in the case of a reflex move by themselves. A good example of a reflex is the flexing of the arm when the hand touches something hot or sharp. Other reflex actions include sneezing, blinking and the multiple adjustments to posture that ensure we stay balanced.

The knee-jerk, or patellar, reflex makes the leg straighten when tapped below the knee. Doctors use this reflex to check that the peripheral nerves and spinal cord are functioning correctly. This is because the nervous control of a reflex does not involve the brain. Instead, a sensory nerve, the spinal cord and a motor nerve create a so-called 'reflex arc', which takes control. A stimulus sends a signal along the sensory nerve to the spinal cord. There, it is redirected to a motor nerve, which orders the correct muscle to contract. The brain is aware of the stimulus and response – which is why it can feel odd – and can override if necessary.

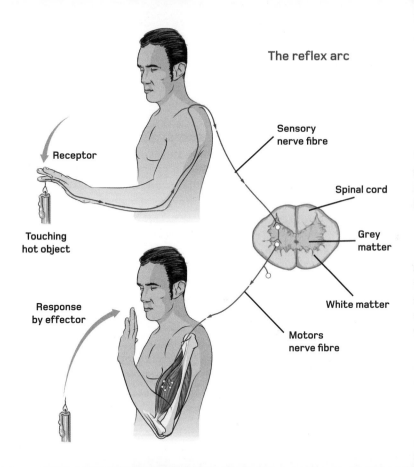

The reflex arc

Receptor

Touching
hot object

Response
by effector

Sensory
nerve fibre

Spinal cord

Grey
matter

White matter

Motors
nerve fibre

The brain

The human brain is probably the most complex object in the Universe. It weighs about 1.5 kilograms (3 lb 5 oz) and is largely made of fat. The best estimate so far is that it contains 83 billion neurons, which together consume 20 per cent of the body's energy supplies. Each one of these neurons is connected to its neighbours. Some may have only a few connections, while others have around 100,000. A conservative estimate puts the number of connections at 100 trillion in all.

These connections are key to the way in which the brain works. Each neuron is constantly making and breaking connections with its neighbours, or as neuroscientists say, 'cells that fire together, wire together'. Quite how the myriad of connections work in order to process information coming from the body and make sense of it – how the brain models the body's surroundings and makes, stores and recalls memories about it – while also conjuring a conscious mind, is still largely a mystery. The question remains: is the human brain clever enough to understand itself?

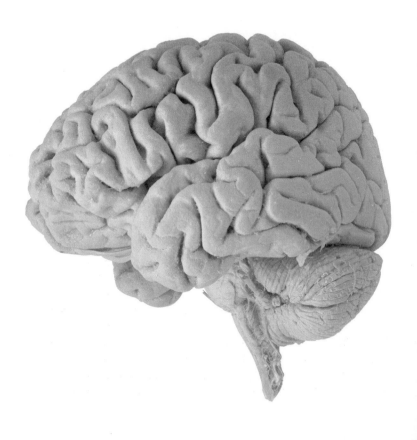

Brain structure

One goal of neuroscience is to plot a functional map of the human brain – in other words, to figure out which part does what exactly, and how it relates to all the other regions. There are several versions of such a map, with functions such as speech, motor control, vision and hearing isolated in certain areas. However, to date, the higher executive functions, such as thinking and remembering, are proving hard to pin down.

The brain has an anatomical hierarchy. The primitive functions are located in the lower parts of the brain, in what's referred to as the limbic system. The hindbrain comprises the medulla oblongata, which forms the connection to the spinal cord and controls breathing and heart rate; the cerebellum, which coordinates muscle control; and the pons, which is involved with swallowing. The midbrain, a small area above, manages body temperature and arousal. The largest part of the brain, the cerebrum, is where the human brain handles sensory perception, where it stores memories and where it thinks, imagines and dreams.

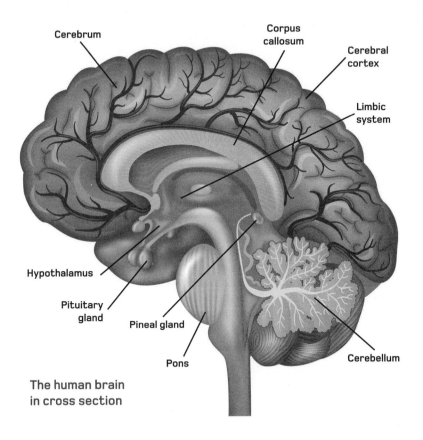

Cerebrum

Corpus
callosum

Cerebral
cortex

Limbic
system

Hypothalamus

Pituitary
gland

Pineal gland

Pons

Cerebellum

The human brain
in cross section

The hypothalamus and limbic system

Two important regions sit at the centre of the brain. The hypothalamus is the primary hormone-producing area, closely associated with the pituitary and pineal glands. Together they produce chemical messengers that manage the adrenal gland, which produces stress hormones; the thyroid, which controls metabolic rate; and the menstrual cycle. It also produces growth hormone. The term hypothalamus means 'under the thalamus', a region of the brain that is part of the limbic system.

The limbic system, surrounding the thalamus, is the seat of our primitive, even animalistic, emotions, such as fear, pleasure and anger, as well as our basic appetites – to eat, sleep and have sex. It communicates with the rest of the body, at least in part, via the hypothalamus. In a healthy brain, its impulses are inhibited by the higher, more 'human', control systems. Without that inhibition, the body flies into what appears to be a rage. Technically, this is a 'sham rage', because without the higher brain, the wild, furious movements lack planning or coordination.

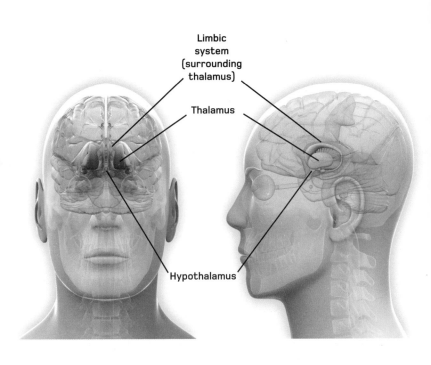

Limbic
system
(surrounding
thalamus)

Thalamus

Hypothalamus

The cerebellum

The cerebellum, meaning 'little brain', is a distinctive structure at the back of the cranium, beneath the cerebrum. It was one of the first regions of the brain to be identified, with Galen, the most influential medical mind of the Roman era, thinking it was some kind of valve that mediated the 'animal spirits' flowing from the brain to the muscles.

This is the fastest-growing part of the brain after birth, developing into one-tenth of the total brain volume by the age of two. This is the period during which an infant is developing motor skills, which might suggest Galen's ancient theory was not entirely off the mark. Research during World War I showed that damage to this region caused movements to be jerky and poorly controlled. The cerebellum stores 'motor schemes' or 'muscle memory'. These are pre-learned, coordinated movements of sets of muscles to make them contract and relax in a smooth, concerted way. A command coming from the higher brain is matched to the most appropriate scheme.

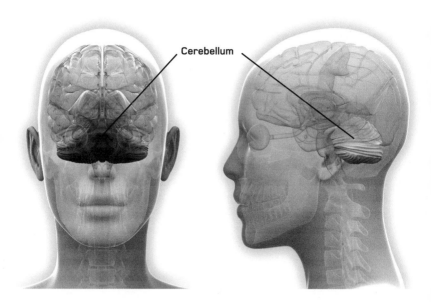

Cerebellum

The cerebrum

The cerebrum makes up 85 per cent of the brain. Its primary region is a thin layer of grey matter called the cerebral cortex. This is folded into ridges (gyri) and grooves (sulci) to increase its surface area. Memory, perception and consciousness take place here. The cells of the cortex connect to white matter deeper inside the cerebrum, which transmit signals to and from the lower parts of the brain.

The cerebrum is divided into two hemispheres, left and right, which are connected at the centre by a bundle of nerves called the corpus callosum. Each hemisphere is divided into four lobes, frontal (front), temporal (side), parietal (top) and occipital (back), each delineated by deep sulci. In very broad terms, the occipital lobe is most associated with vision. The temporal lobe manages hearing and processes other sensory information and associates that with memories. The parietal lobe works with spatial awareness and body position, while the frontal lobe controls movements and the executive functions.

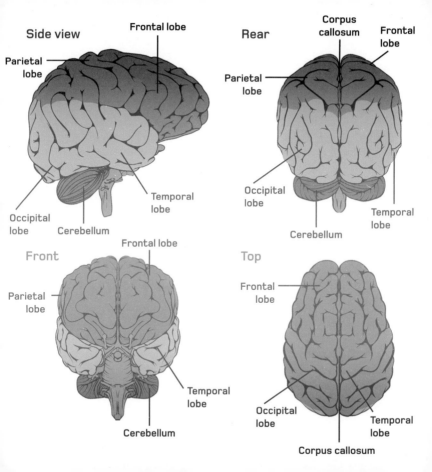

Cranial nerves

Most of the work of the brain is mediated through the spinal cord and peripheral nervous system beyond it. However, the main sense organs on the head and the muscles of the mouth, throat, face and eyes are connected directly to the brain by 12 sets of cranial nerves. These nerves are the reason why a spinal injury that severs the spinal cord may lead to paralysis below the neck but does not paralyse the head.

Most cranial nerves have either a sensory or a motor function. For example, the ears, nose and eyes have one sensory nerve each devoted to them. (The eye also has three motor nerves to control movement.) A few sets of nerves, such as nerve IX, are afferent and efferent pairs – one carries taste information, the other controls the throat and salivary glands. The longest cranial nerves are the vagus nerves (another paired set), which travel through the neck all the way to the top of the colon. Their many crucial roles include controlling the vocal cords and gag reflex, heart rate, and the movements of the digestive system.

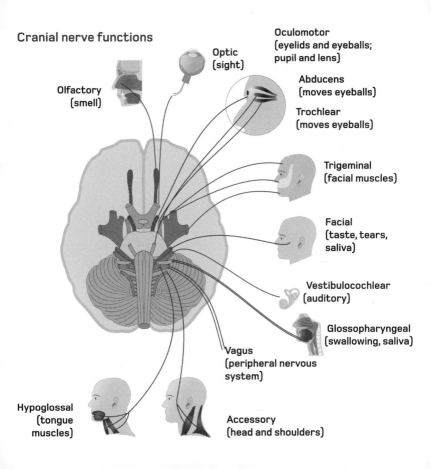

Cranial nerve functions

Olfactory (smell)

Optic (sight)

Oculomotor (eyelids and eyeballs; pupil and lens)

Abducens (moves eyeballs)

Trochlear (moves eyeballs)

Trigeminal (facial muscles)

Facial (taste, tears, saliva)

Vestibulocochlear (auditory)

Glossopharyngeal (swallowing, saliva)

Vagus (peripheral nervous system)

Hypoglossal (tongue muscles)

Accessory (head and shoulders)

The eye

The human eye is an elongated ovoid about 2.5 centimetres (1 in) across at its widest point. It is about 60 per cent of its adult size at birth and reaches full size at the age of 13.

Most of the outside of the eye is covered in a white sclera. The front is a transparent bulge called the cornea, under which is the iris. This is a muscular ring that forms the aperture through which light enters. It can open and close to control how much light gets in. Behind the iris is the lens. This is a flexible capsule that can be stretched to focus light on to the retina on the back of the eye. (The cornea also helps with focusing.) All 'empty' space in the eye is filled with a transparent gel-like liquid., the vitreous humour. The retina is a layer of cells containing light-sensitive chemicals that trigger nerve impulses when excited by light. Three types of cone cell in the retina – mostly around the central focus point – each detect a colour, either red, green or blue. The more sensitive rod cells are used for seeing in dark conditions, but cannot detect colour and so see in black and white.

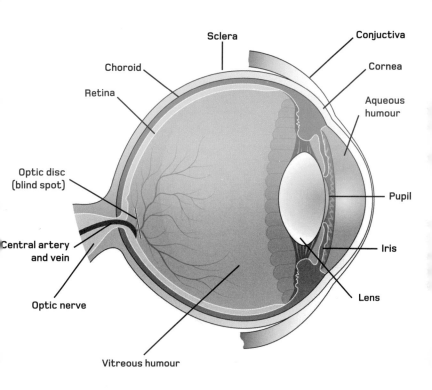

Vision

The chemical response to light inside retinal cells is converted to nerve impulses. These travel through small nerve cells that actually cover the retina (light shines through them) and converge at the optic nerve, which passes through the retina and out of the eye.

The nerves from each eye meet at a junction called the optic chiasma. Here, the signals are reorganized. Information from the part of each retina that looks left is sent to the right occipital lobe, while information concerning the view to the right is sent to the left side of the brain. Along the way, the information passes through the thalamus. Some information is redirected to the superior colliculus in the midbrain, which is involved in coordinating eye movements. The main optic nerves continue to the visual cortex in the occipital lobe. The pattern of light and dark as recorded by the retinal cells is then mapped on to cells of the cortex. The regions neighbouring the cortex then associate that image with previous visual memories.

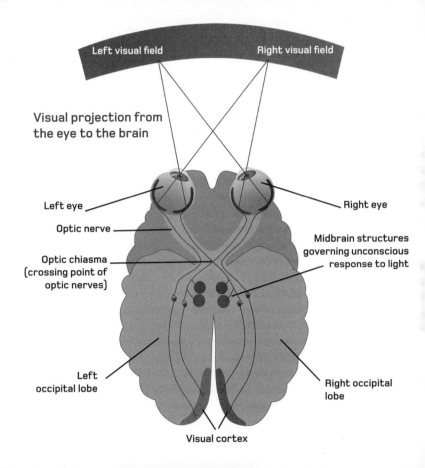

Left visual field
Right visual field

Visual projection from
the eye to the brain

Left eye
Right eye

Optic nerve

Optic chiasma
(crossing point of
optic nerves)

Midbrain structures
governing unconscious
response to light

Left
occipital
lobe

Right occipital
lobe

Visual cortex

The ear

The human ear is in three parts: outer, middle and inner. The outer ear consists of the pinna and the ear canal. The pinna is a spiral of cartilage that creates a dish-shaped structure to collect sound waves and focus them into the opening of the ear canal. The canal is 2–3 centimetres (³⁄₄–1¹⁄₄ in) long and ends with the eardrum, a flap of skin that covers the canal and that vibrates when sound waves hit it.

The middle ear is behind the eardrum. It is an air-filled cavity that connects to the outside via a small channel, called the Eustachian tube, which runs into the back of the throat. This tube allows air pressure either side of the eardrum to equalize if necessary – making your ears 'pop'. The middle ear holds three tiny bones, or ossicles, known as the hammer, anvil and stirrup.

The inner ear is a convoluted bone structure filled with fluid, called the labyrinth. This part of the ear is involved in maintaining balance as well as hearing.

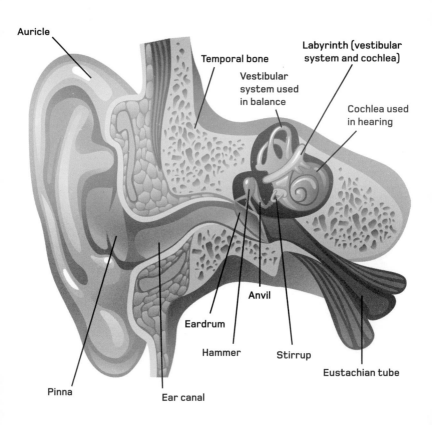

Hearing

The ear converts the energy in sound waves into nerve impulses using a series of steps. First, the sound is focused into the ear canal by the dish-shaped pinna. The wave hits the eardrum, a layer of skin across the canal, making it vibrate with matching rhythm. The next step takes place in the middle ear, where the wave is further transmitted via the ossicles.

The malleus (hammer bone) is connected to the eardrum. It taps a rhythmic signal on the incus (anvil), which passes it on to the curved stapes (stirrup bone). The stapes sits on the surface of the fluid-filled labyrinth situated in the inner ear. A coiled section of the labyrinth called the cochlea is concerned with hearing. The taps on the stapes pass through a membrane-covered 'oval window' and set up a corresponding wave in the fluid inside the cochlea. This wave travels to the organ of Corti deeper in the cochlea, which is lined with ciliated cells (hairlike structures). The wave wafts the cilia, stimulating the cells to send signals along the auditory nerve to the brain.

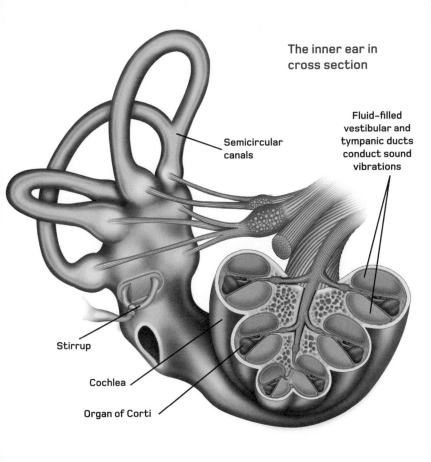

The inner ear in cross section

Semicircular canals

Fluid-filled vestibular and tympanic ducts conduct sound vibrations

Stirrup

Cochlea

Organ of Corti

Sense of balance

The labyrinth of the inner ear (see page 170) is also used to keep the body balanced in a sensory process called equilibrioception. The eyes, and information from the muscles about body posture, also help to keep the body balanced, but the main input comes from the vestibular system.

This is a set of three looped canals located above the cochlea. They are orientated roughly at 90 degrees to one another, so each one detects motion in one of three planes. When the head moves in any direction, the inertia of the fluid inside the vestibular system creates an apparent flow along the relevant canal. This is detected by lumps in the canals called the crista ampullaris, which send the information to the brain via the vestibulocochlear nerve – the same one that carries auditory signals. Any motion detected by the vestibular system is translated by a reflex action to the eyes, to ensure that they maintain a fixed gaze. Information from the vestibular system overrides conflicting spatial information from other senses. When it cannot do this, you feel dizzy.

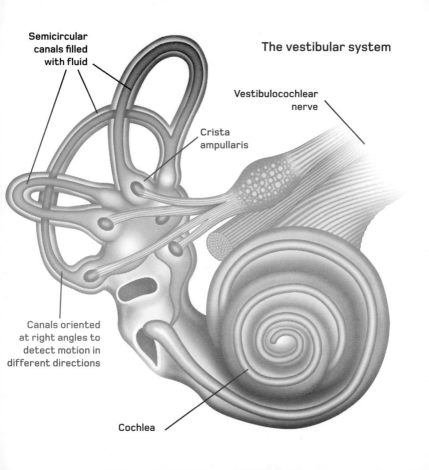

Semicircular canals filled with fluid

The vestibular system

Vestibulocochlear nerve

Crista ampullaris

Canals oriented at right angles to detect motion in different directions

Cochlea

The nose

The sense of smell, or olfaction, is the detection of chemicals in the air. This is achieved using a thumb-sized patch of cells located in the nasal cavity, forming what is called the olfactory epithelium. This tissue is made up of many hundreds of different kinds of chemoreceptor cells, which detect odour chemicals dissolving in the mucus that lines the nasal cavity.

The receptor cells have several types of ion channels on their surface. Each channel opens when a specific odour chemical binds to it. The ensuing flow of ions initiates an action potential that leads to a nerve signal heading to the brain. Nerve endings from each receptor pass through the skull to a mass of nerve fibres called the olfactory bulb. From there, olfactory impulses travel through part of the limbic system (see page 156), before reaching the olfactory cortex at the front of the brain. The receptor cells in the human nose can detect 10,000 unique chemicals. These chemicals blend together creating millions of distinctive smells. Women have a more sensitive sense of smell than men.

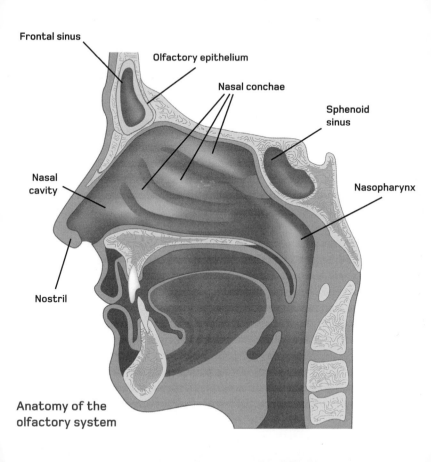

Frontal sinus

Olfactory epithelium

Nasal conchae

Sphenoid sinus

Nasal cavity

Nasopharynx

Nostril

Anatomy of the
olfactory system

The tongue

The tongue is a slab of muscle in the mouth, used to manipulate and taste food. It is made up of eight muscles. Four float free of the skeleton and are used to alter the shape of the tongue. The others connect to the jaw, skull and other anchor points to move the tongue in the mouth during licking, chewing, swallowing and talking.

The tongue is the primary taste organ, although the lining of the mouth and throat can pick up flavours, too. Gustation, or the sense of taste, is closely allied to the sense of smell. We even smell our food as we chew it, which is why we cannot taste well with a blocked nose. The tongue is covered in bumps, or papillae, which contain clusters of taste buds. These hollow, bulb-shaped pits are filled with chemoreceptor cells. The cells can detect at least five flavours: sweet, salty, bitter, sour and umami, or 'meatiness', although the idea that a different region of the tongue is devoted to detecting each flavour is a misconception. Recently, a sixth starchy flavour has been proposed.

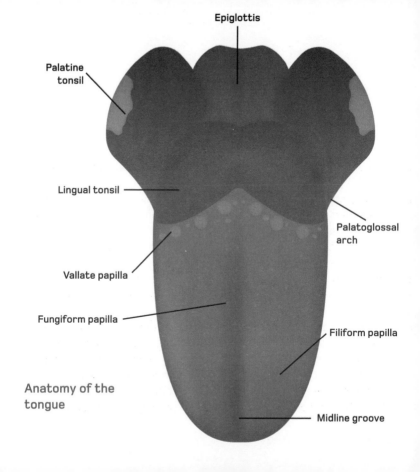

Anatomy of the
tongue

The autonomic nervous system

The unconscious control of body systems is handled by the autonomic nervous system. Using a different set of nerves to the somatic nervous system, this divides into two subsystems.

The sympathetic nervous system connects to all major organs using nerves that emanate from two chains of nerve bundles, or ganglia, which run down each side of the spinal cord. This system stimulates the stress-related 'fight or flight' functions of the body. It elevates heartbeat, enlarges the lungs, inhibits digestion and urination, and releases energy stores from the liver. It also stimulates the genitals during sexual arousal.

By contrast, the parasympathetic system does the opposite – the so-called 'rest and digest' functions. Its nerves mostly form a direct link between the brain and the organs, although the bowels, bladder and genitals are controlled by sacral nerves at the tip of the spinal cord. They reduce the heart and breathing rate and divert energy to the digestive system.

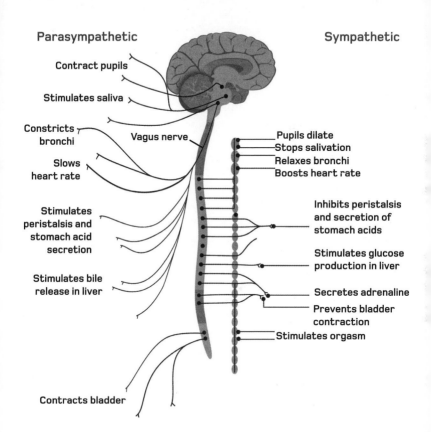

Parasympathetic

Sympathetic

Contract pupils

Stimulates saliva

Constricts bronchi

Slows heart rate

Vagus nerve

Stimulates peristalsis and stomach acid secretion

Stimulates bile release in liver

Contracts bladder

Pupils dilate
Stops salivation
Relaxes bronchi
Boosts heart rate

Inhibits peristalsis and secretion of stomach acids

Stimulates glucose production in liver

Secretes adrenaline

Prevents bladder contraction

Stimulates orgasm

The skin

The skin is a complex of epithelial, connective and sensory tissues that along with other materials, such as hair and nails, make up the integumentary system. It is sometimes said that skin is the largest organ in the body, because it covers approximately 1.8 square metres (19 ft^2) when spread flat. However, both the lungs and the intestines have significantly larger surface areas when their fine structures are accounted for. Nevertheless, the skin is a very important body part.

Skin has three layers. The epidermis is the outer one, then the dermis and finally the hypodermis – a layer of fat that connects to the muscles, bones and organs beneath. The skin has multiple roles: it forms a waterproof barrier that stops body fluids leaking out and foreign material getting in. It is also involved in temperature regulation through sweating and other mechanisms. Finally, it is the location of the great majority of the somatosensory organs. Often referred to as touch sensors, these are capable of much more than that implies.

Structure of human skin

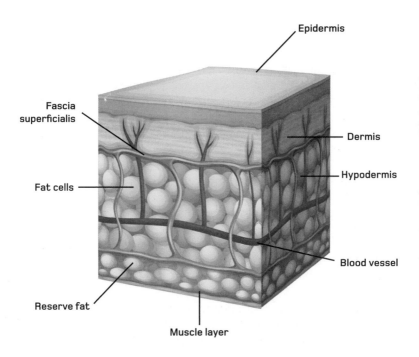

Epidermis

Fascia superficialis

Dermis

Hypodermis

Fat cells

Blood vessel

Reserve fat

Muscle layer

The epidermis

The outermost layer of the skin is the epidermis. In terms of tissue type, this is squamous epithelium, made up of row upon row of cells that are wider than they are deep. There is no blood supply in the epidermis. The cells grow from a base layer and survive by oxygen diffusing in from the air – the skin really does 'breathe'.

As new layers are added underneath, the cells above begin to die. The uppermost layers are made entirely of dead cells, which slough off slowly but surely. This outward growth and resulting loss of epidermal cells is an active process called desquamation that also removes any foreign objects that have become attached. The entire surface of the body is renewed in this way every month.

The top 20 or 30 layers of epidermal cells have no cytoplasm or active cell contents. In their place is a sac of keratin, a waxy, protein that makes the epidermis waterproof.

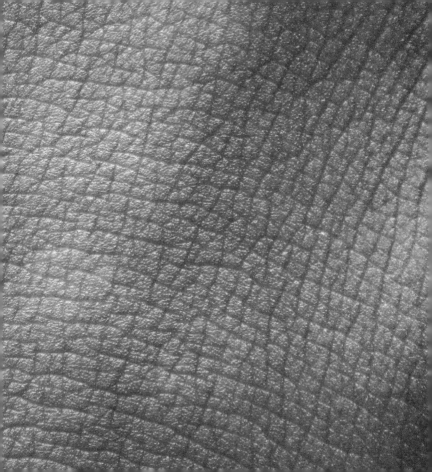

The dermis and hypodermis

The dermis is a connective tissue that bonds to the epidermis above with a basement membrane made of collagen. It is four or five times thicker than the epidermis.

Made up of fibrous and elastic tissues, the dermis has strong but flexible properties. It contains blood and lymph vessels and is rich in white blood cells, ever ready to dispose of invaders and aid with clotting and healing in the event of an injury. If a wound to the skin cuts through its basement membrane, a scar will form after healing. The dermis also houses most of the receptors of the somatosensory system, although some are located in the next layer down, the hypodermis. This deepest layer is the part of the integumentary system that connects to the subcutaneous structures – muscles, bone, and so on – but is not strictly a skin tissue. It is mostly adipose tissue, where fat is stored. This fat layer is thickest under the less hairy parts of the skin (back, belly and buttocks), and extends deeper into the body to surround and support organs.

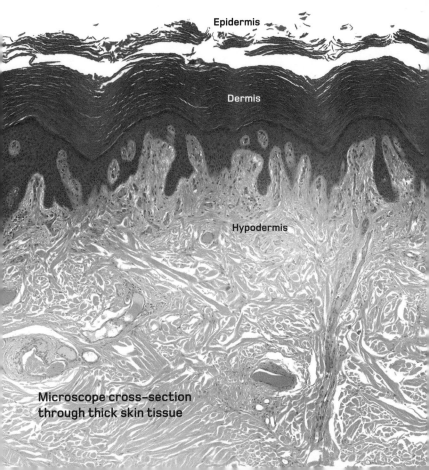

Epidermis

Dermis

Hypodermis

Microscope cross-section
through thick skin tissue

Nails

The nail is a hardened covering on the tips of the fingers, thumbs and toes. The hard visible part is the nail plate, or body, which is made from compact layers of dead keratinized cells. These grow from the nail matrix, or root, which is mostly hidden under the skin, although its moon-shaped tip, the lunula, is visible at the base of the plate.

The nail plate is attached to a curved layer of skin tissue beneath, called the nail bed. The bed's epidermal layer migrates towards the fingertip as the nail grows. Nails grow at about 3 millimetres a month, although fingernails grow faster than toenails. In developmental terms, the human nail is an unguis, the analogue of a claw or hoof. Most monkeys and all apes have nails, but they are rare among other small and medium-sized mammals. They usually sport claws that are laterally compressed and pointed. The reason why we have nails instead is not fully clear. It is likely our ancestors evolved them as their fingertips grew wide and fleshy with ridges for gripping small branches.

Nail anatomy

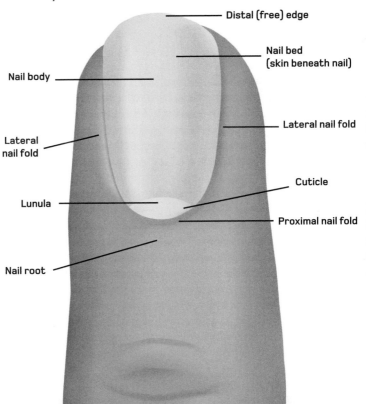

Distal (free) edge

Nail bed
(skin beneath nail)

Nail body

Lateral nail fold

Lateral
nail fold

Cuticle

Lunula

Proximal nail fold

Nail root

Hair

The human being is the 'naked ape', yet the human body has the same density of hair as that of its ape cousin. The hairs are simply shorter and finer. Why humans have only a thin covering of hair is an open question. It may have helped with temperature control in our evolutionary past, with our heads staying hairy to insulate the brain from the baking sun. Or perhaps it is down to sexual selection, with the loss of thick hair on the body highlighting our sexual characteristics more.

A hair is a shaft of keratin, and other proteins, that grows from a follicle in the dermis. It is structured in three concentric layers. The central medulla is loosely structured and may be hollow. The cortex is packed with keratin and contains the pigments that give the hair its colour. The outer covering is the cuticle. Body hair is used in reducing the loss of body heat. Each follicle is teamed with a tiny muscle that raises the hair when it contracts – making a goosebump on the skin. The erect hairs trap an insulating layer of stationary air around the skin.

Structure and embedding of hair

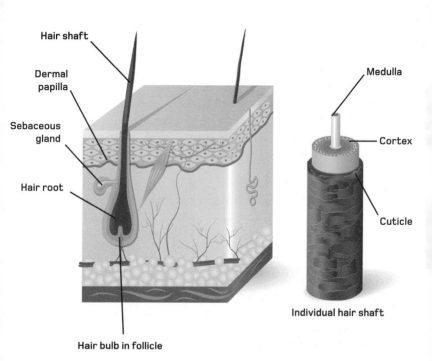

Hair shaft

Dermal papilla

Sebaceous gland

Hair root

Hair bulb in follicle

Medulla

Cortex

Cuticle

Individual hair shaft

Hair colour

Human hair is incredibly diverse. Along with face shape, it is one of the ways in which we identify each other. A hair's breadth can range from about 0.2 to 0.02 millimetres, and it can be long and straight or short and curly, depending on the follicle shape. Great attention is given to the colour of hair – with vast sums spent on dyes to change or retain it.

Human hair colour is controlled by two types of melanin, or pigment chemicals, called eumelanin and pheomelanin. Black hair is the most common colour, and contains mostly eumelanin. It can even have hints of blue. Red hair, which has mostly pheomelanin, is the rarest colour; outside of northwestern Europe and North America, where about five per cent of people are redheads, the frequency drops to less than one per cent. Blonde hair contains both melanins, but only in small quantities. Brown hair also has a mix of both but in larger quantities, thus forming a continuum of shades between the other three. Grey hair appears as the melanins begin to disappear; white hair has no melanin at all.

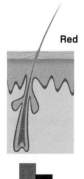

Red

Pheomelanin dominates

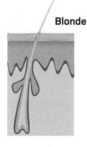

Blonde

Small amounts of both pigments

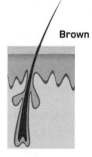

Brown

Larger amounts of both pigments

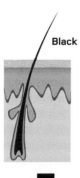

Black

Eumelanin dominates

Grey

Pigments diminished

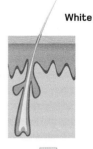

White

Pigments replaced by air

Eyelashes and eyebrows

The human eye has two sets of hairs with specific functions. The eyebrows, known in anatomical parlance as the supercilia, grow on the fleshy ridge above the eye. It is assumed that their main function is to keep sweat (and perhaps rain) from dripping into the eyes. The arc of hair redirects the water to the sides or accumulates it for wiping away periodically.

Another probable function of the eyebrows in humans is in communication. The nuances of facial expressions are enhanced by the motion of the eyebrows. The whites of the eye, a human-only trait, also help here.

The eyelashes, or cilia, have a clearer function. These thick hairs grow from the edges of the eyelid. They are highly sensitive to contact and stimulate the eye to close rapidly when something touches them. The eyelashes are longer and bushier on the upper eyelid, where they form a brush-like fringe that helps to shield the eye from grit or other wind-blown irritants.

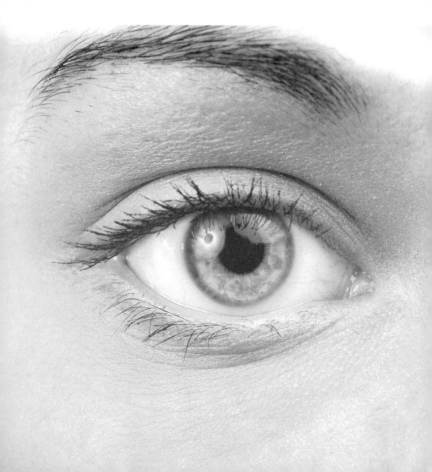

Sweating

The human skin has two million sweat glands on average (the number varies widely for different people). They are most concentrated in the palms of the hand and are closely packed in the groin and the armpit. The back and legs are the least sweaty regions with six times fewer sweat glands than the hands. Only the lips lack sweat glands altogether.

Sweat glands are tubular structures in the hypodermis (see page 184). The great majority of human sweat is made by eccrine glands. These produce salty water, which is pumped up the tube and out of a pore in the skin by contracting the cells around it. This salty, slightly acidic sweat is used for cooling. As the liquid evaporates away from the skin, it carries some heat energy with it. The sweat is also being used to get rid of excess sodium ions and tiny amounts of urea, the nitrogenous waste normally expelled in urine. The sweat glands in the armpit, groin, ear canal and around the eyes are apocrine glands. These produce oilier, more alkaline sweat.

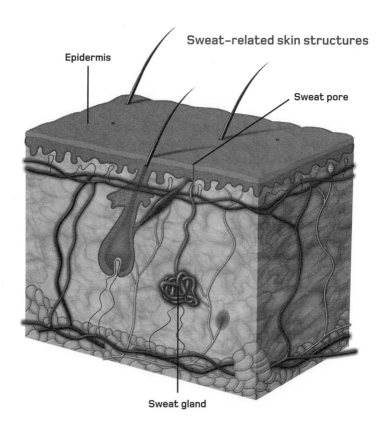

Sweat-related skin structures

Epidermis

Sweat pore

Sweat gland

Sense of touch

The sense of touch is also known as the somatosensory system. Sensors are mostly in the skin, but are also present in muscles, bones and many other body systems.

There are six receptors at work; the first four are mechanical and respond to a physical deformation. Meissner corpuscles, located in the upper dermis, are fast acting and detect light pressure and so create the fine touch sense. Pacinian corpuscles are similar, but deeper down, and so detect harder and rougher contacts. The Merkel discs at the bottom of the epidermis are slower-firing and pick up persistent contact, while the Ruffini corpuscles are spread throughout the dermis and respond to stretching. They give the sense of an object moving over the skin. Krause's corpuscles also in the upper dermis, and are responsible for detecting cold – how they do it remains something of a mystery. Meanwhile, heat is generally picked up by free nerve endings under the epidermis. These are also involved in sending the signals that lead to a pain sensation.

Touch sensors in the skin

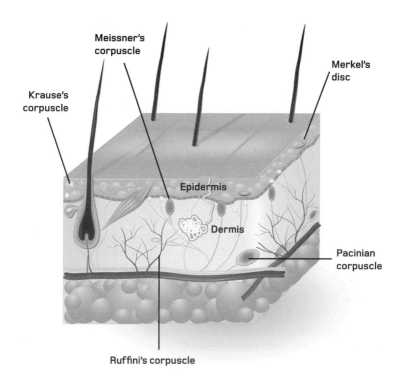

Meissner's corpuscle

Merkel's disc

Krause's corpuscle

Epidermis

Dermis

Pacinian corpuscle

Ruffini's corpuscle

Somatosensory cortex

Every inch of skin is part of a dermatome, or an 'island of sensibility'. Each dermatome has its own sensory nerve carrying touch information to the brain. However, dermatomes are not all the same size. Those on the back are much larger, and therefore less able to resolve touch information as well as a smaller dermatome, such as on the fingertips or genitals. There are similar somatosensory units that allow us to feel inside the body as well.

Information from each dermatome travels via the spinal cord to the somatosensory cortex in the parietal lobe. Here, sensations from the body are mapped onto specific regions of the cortex. The more sensitive regions, such as the face and tongue, have more brain space allocated to them than the trunk or limbs. Next to the somatosensory cortex, over in the frontal lobe, is the motor cortex. This has a similar body map but organizes its space differently, by devoting space according to those body parts that require the finest motor controls.

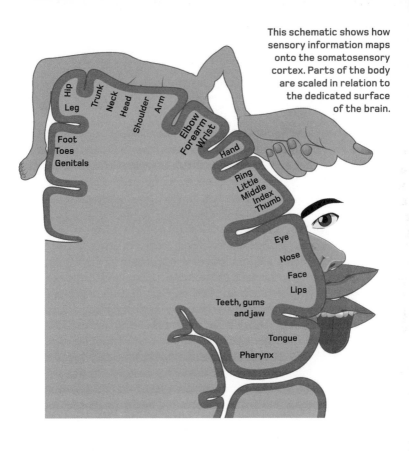

This schematic shows how sensory information maps onto the somatosensory cortex. Parts of the body are scaled in relation to the dedicated surface of the brain.

Hip
Leg
Trunk
Neck
Head
Shoulder
Arm
Elbow
Forearm
Wrist
Hand
Foot
Toes
Genitals
Ring
Little
Middle
Index
Thumb
Eye
Nose
Face
Lips
Teeth, gums and jaw
Tongue
Pharynx

Pain

Pain is the body's way of warning that tissue has been damaged or is likely to become damaged. The body responds, often by reflex, to remove itself from the source of the problem.

Pain is the result of signals from nociceptors. These produce three types of pain: thermal, mechanical and chemical. The first results from receptor chemicals in nerve endings (in the skin, mostly) rising above 42°C (108°F), which stimulate a nerve signal. Mechanical pain is the result of nerve endings being physically deformed. Lastly, certain chemicals can cause a receptor to fire. (Chilli does this by triggering pain receptors in the mouth; the taste buds are not involved.) After an injury the area becomes inflamed, increasing the sensitivity of the nociceptors as a means of protecting the body as it heals. Much of the body's response to pain is involuntary, involving reflexes and immune responses. Why then do we feel the distress of pain? Is the sensation of pain an artefact of human consciousness, and if so, what does it tell us about the mental states of other animals?

Transfer of pain stimuli to the brain

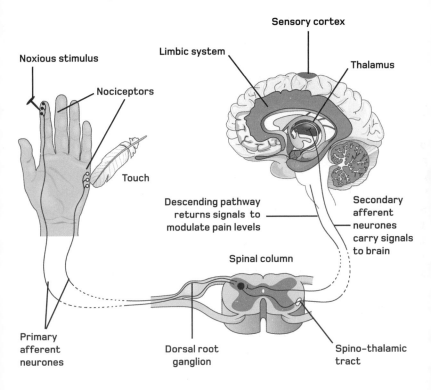

Bodily secretions

The body is not controlled solely by electrical impulses transmitted via the nervous system (see page 140). There is another set of control mechanisms, and this involves chemical messages called hormones.

These are secreted from cells most often located in body parts called glands, which release hormones into the blood stream or a major organ. Together, these glands make up the endocrine system. The method used by these secretory cells to release their secretions is called exocytosis. Within the cell, the hormone or enzyme is parcelled into a vesicle – a bag surrounded by a membrane. This membrane merges with the cell's outer membrane and the contents emerge from the cell.

There are other glands at work. These secrete chemicals out of the body, generally via ducts that pass through the skin, or into the digestive tract. These are exocrine glands, and produce materials such as saliva, mucus, stomach enzymes and sweat.

Exocytosis

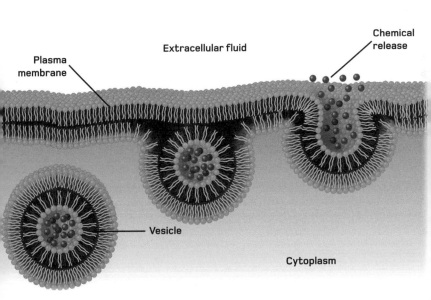

Chemical release

Plasma membrane

Extracellular fluid

Vesicle

Cytoplasm

Glands

There are many millions of exocrine glands all over the body surface and digestive tract, and they fall into a number of broad categories. As you might expect, mucus glands, such as goblet cells, produce gooey mucus. Serous glands produce watery secretions like saliva and sweat, and sebaceous glands produce oily products, such as sebum, which keeps skin supple.

By contrast, the endocrine system is dominated by a series of large, centralized glands – eight in all. The pineal gland, pituitary gland and hypothalamus are all in the brain, although the first two are not nervous tissue. The pineal body produces the sleep hormone, melatonin.

From the neck down the glands are: the thyroid; the thymus, which is involved in immunity and becomes inactive after puberty; the pancreas, which works as an exocrine gland as well; a pair of adrenal glands, each one on top of a kidney; and, finally, the gonads, or sex glands, of which there are also two.

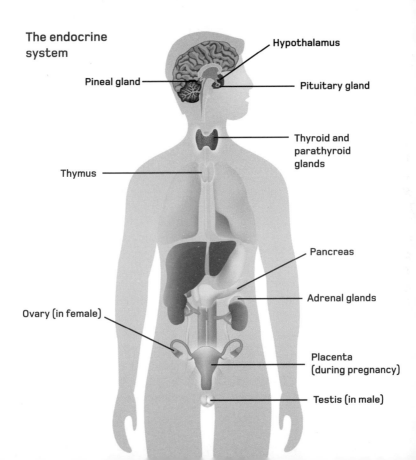

The endocrine system

Hypothalamus

Pineal gland

Pituitary gland

Thyroid and parathyroid glands

Thymus

Pancreas

Adrenal glands

Ovary (in female)

Placenta (during pregnancy)

Testis (in male)

Sebum

Sebum is an oily or waxy secretion that is released on to the skin. Its function is to keep the skin and hair moisturized and to make it waterproof. It also mixes with sweat to make it thicker and more cohesive, thus forming a cooling film over the skin.

Sebum is produced by sebaceous glands in the upper region of the dermis. Each one is associated with a hair, and deposits its sebum into the follicle, from where it moves to the skin. The glands are most concentrated on the face, palms and soles of the feet. Specialized sebaceous glands in the eyelids mix with tears to form a protective coating over the eyeball. Earwax, or cerumen, is also produced by these glands in the ear canal. The wax travels down the canal removing dirt with it. Sebaceous glands do not release sebum by exocytosis. Instead, they are holocrine glands, which means the cells fill up with sebum and then disintegrate, releasing their contents. If a hair follicle becomes blocked by sebum, the secretions build up behind it and may become infected, forming a pimple or spot.

How a spot forms

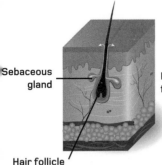

Sebaceous gland

Hair follicle

Healthy follicle

Sebaceous duct becomes clogged and sebum begins to accumulate.

Bacterial infection leads to inflammation and a pimple starts to form.

Follicle ruptures and a pustule filled with fluid forms.

Tears

Tears are a liquid secretion onto the eyes produced by endocrine lacrimal glands. There are two main clusters of these, one above the eye on the outer side, and another between the eye and the bridge of the nose. These secrete tears onto the surface of the eyeball via series of ducts around the eye. The largest duct is in the inner corner of the eye, next to the nose.

Tears keep the surface of the eye moist and clean, a process achieved mainly with basal tears, which are thickened slightly with mucus and oils. Produced by the superior gland above the eye, they are wiped over the surface by frequent blinking. When the eye becomes irritated, the glands produce more watery tears in an attempt to wash away the problem. Tears are also produced when a person is emotional – mostly sad or in pain. These 'psychic' tears contain stress hormones and pain suppressants. Crying may have some physiological role in excreting stress-related chemicals, and acts as an emotional release allowing the mood to return to a more normal state of arousal.

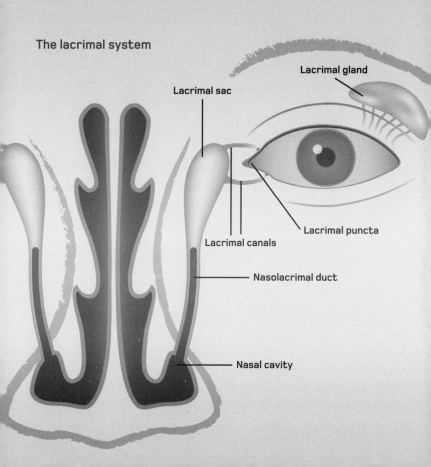

The lacrimal system

Lacrimal gland

Lacrimal sac

Lacrimal puncta

Lacrimal canals

Nasolacrimal duct

Nasal cavity

Mucus

Mucus is a slippery liquid secretion used to lubricate and protect the lining of body cavities, such as the digestive tract, lungs and vagina. It is a saturated mixture of water and large proteins called mucins. When mixed, they create a gel.

Mucus is largely produced by goblet cells, which are column-shaped cells located in large numbers within the epithelial tissues. Goblet cells are effectively single-celled glands. They are divided in two halves. The nucleus and other generalized cell machinery are located at the bottom, away from the body cavity surface. The upper part of the cell is filled with endoplasmic reticulum and a large Golgi apparatus. The reticulum – a network of membranous tubes – manufactures the mucins, and then the Golgi apparatus parcels them up with water into vesicles. The vesicles migrate to the top end of the cell and then leave by exocytosis. The mucus might be delivered to a duct that feeds a tissue surface, or it might be delivered directly to where it is needed.

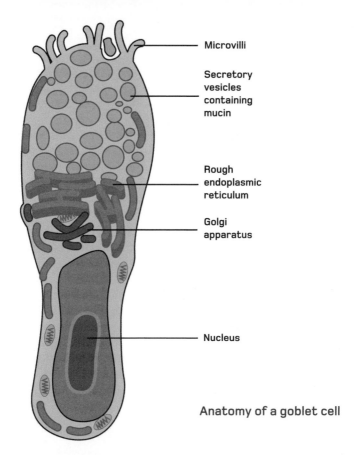

Microvilli

Secretory vesicles containing mucin

Rough endoplasmic reticulum

Golgi apparatus

Nucleus

Anatomy of a goblet cell

Hormones

Organs and body systems orchestrate their activities using hormones. Produced by glands around the body, hormones are slower-acting than nervous signals.

Hormones have different effects on different body parts – stimulating one and inhibiting the actions of another. They do not create 'all or nothing' processes like nervous signals. Instead, their impact is generally controlled by feedback systems: the action of one hormone effects the secretion and action of another. As a result, different hormone levels rise and fall in response to specific conditions, and generally oscillate around an optimal level that keeps the body's processes running smoothly. There are three main types of hormone chemicals. Steroids are modified cholesterol molecules. They include the sex hormone, testosterone. Eicosanoids are based on large fat molecules and include prostaglandins, which are involved in inflammation. The third type, amino acids (the subunits of proteins), include adrenalin and insulin.

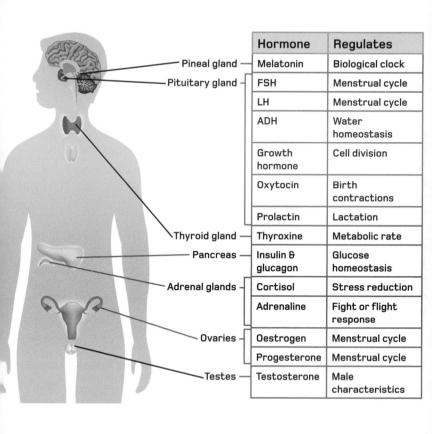

Hormone	Regulates
Melatonin	Biological clock
FSH	Menstrual cycle
LH	Menstrual cycle
ADH	Water homeostasis
Growth hormone	Cell division
Oxytocin	Birth contractions
Prolactin	Lactation
Thyroxine	Metabolic rate
Insulin & glucagon	Glucose homeostasis
Cortisol	Stress reduction
Adrenaline	Fight or flight response
Oestrogen	Menstrual cycle
Progesterone	Menstrual cycle
Testosterone	Male characteristics

Pineal gland

Pituitary gland

Thyroid gland

Pancreas

Adrenal glands

Ovaries

Testes

The pancreas

The pancreas is the largest gland in the human body, located behind the lower part of the stomach. It connects to the duodenum section of the small intestine, where it works as an exocrine gland, secreting digestive enzymes, as well as several hormones used by the endocrine system.

The most significant pancreatic hormones are insulin and its partner, glucagon, both of which are produced in cell clusters called the islets of Langerhans. Insulin stimulates body tissues – mostly in the liver – to absorb the sugar released into the blood by digestion. The absorbed sugar is converted to glycogen, a complex carbohydrate.

High levels of insulin reduce blood sugar. In response, the pancreas produces glucagon, which has the opposite effect. Stored glycogen is converted to sugar and released into the blood. The two hormones work together to maintain a reasonably constant supply of sugar for the body's needs.

The pancreas in cross section

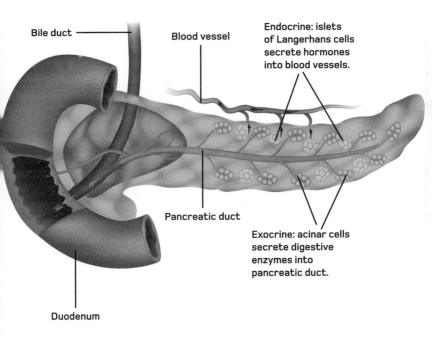

Bile duct

Blood vessel

Endocrine: islets of Langerhans cells secrete hormones into blood vessels.

Pancreatic duct

Exocrine: acinar cells secrete digestive enzymes into pancreatic duct.

Duodenum

The thyroid gland

The thyroid gland is a butterfly-shaped organ wrapped around the windpipe, or trachea, just in front of the larynx. The thyroid is the throttle of the body. It produces a set of hormones, known collectively as thyroxines.

These hormones control the body's basal rate, which is the amount of energy it uses when at rest. They also impact on the rate of protein synthesis, the process where a cell translates its genetic information into useful proteins. Four patches on the larger gland are made of a distinctive tissue and are known as the parathyroid glands. These work separately from the rest of the tissue and produce calcitonin, which controls how much calcium the body absorbs and excretes from the skeleton.

An underactive thyroid (hypothyroidism) makes a person feel lacking in energy, while an overactive one (hyperthyroidism) makes a person have trouble sleeping, lose weight and suffer immunity problems. Both conditions are treatable with drugs.

Anatomy of the thyroid gland

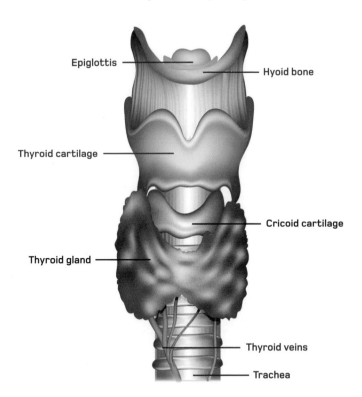

Epiglottis

Hyoid bone

Thyroid cartilage

Cricoid cartilage

Thyroid gland

Thyroid veins

Trachea

The pituitary gland

The pituitary gland works in concert with the hypothalamus (see page 156), a gland-like region of the brain. Despite being about the size of a pea, the pituitary exercises a degree of command and control over the endocrine system by producing hormones that moderate the behaviour of the other glands.

The hypothalamus is associated with emotions and these are translated into physical responses via hormones. The pituitary hormone adrenocorticotrophic hormone (ACTH) is released when the body is stressed, and stimulates the adrenal gland to release cortisol. The pituitary also releases thyroid stimulating hormone (TSH), which causes the thyroid to elevate the baseline metabolic rate. Pituitary hormones are involved in controlling menstruation and lactation and the functions of sex organs. The pituitary also controls water levels in the body via the action of the kidneys. Finally, the pituitary releases growth hormone during childhood, which lengthens bones and enlarges muscles.

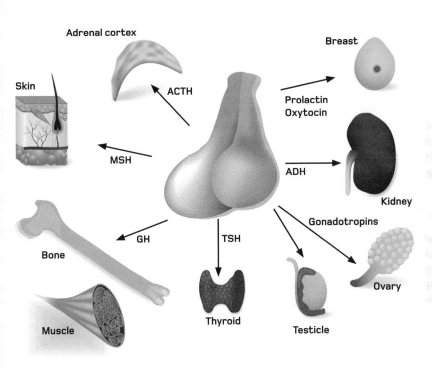

Pituitary hormone functions

The adrenal glands

The body has two adrenal glands, one located above each of the kidneys. They release hormones into the inferior vena cava, one of the largest blood vessels, from where they spread quickly through the body. Their role is to prepare the body for short-term dangers and stressful conditions in the longer term.

The former function is performed by the hormone adrenalin, also known as epinephrine, which is produced by the gland's central region, or medulla. In association with the sympathetic nervous system, this hormone is involved in producing the fight-or-flight response, where the body's resources are diverted to surviving attack. Long-term, low-level stress, such as lack of food and rest, is handled using the hormone cortisol, which is produced by the gland's outer cortex. Cortisol acts to divert limited resources to the survival of the body. It increases blood pressure and blood sugar, slows growth and the maintenance of bones, and suppresses the immune system. This is what creates the link between long-term stress and poor health.

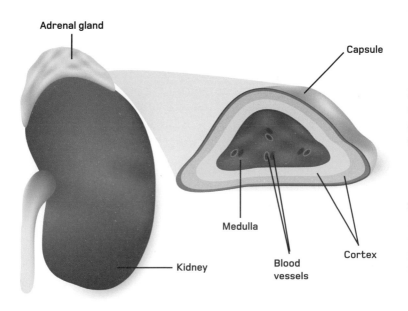

The adrenal gland in cross section

Fight or flight

When danger appears, the body responds very quickly, preparing to escape or fend off the attacker. The body floods with adrenalin made by the adrenal gland and also by parts of the sympathetic nervous system. This hormone acts rapidly on several body systems at once, creating a state of hyperarousal.

The heart rate spikes and breathing rate increases. The blood supply to the digestive system is reduced (creating a sinking feeling) and diverted to the skeletal muscles. The liver begins to release extra sugar into the blood. Meanwhile, the blood supply concentrates in the core body and the skin pales as the veins beneath constrict. The salivary glands also constrict and the mouth dries out. To conserve body heat, the hairs on the skin become erect. Finally, the iris of the eye expands to create a wider field of vision. This process is not foolproof: pumped muscles twitch, making fine motor control difficult; smooth muscles may relax so much that the bowels and bladder empty; and breathing may become too fast and too shallow to collect sufficient oxygen.

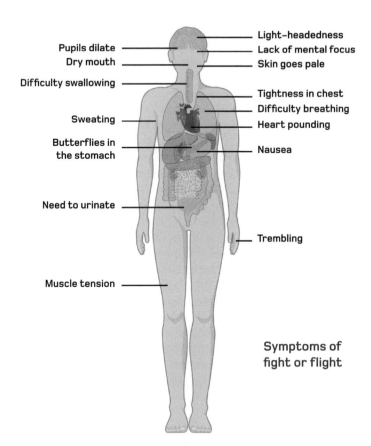

Pupils dilate

Dry mouth

Difficulty swallowing

Sweating

Butterflies in
the stomach

Need to urinate

Muscle tension

Light-headedness

Lack of mental focus

Skin goes pale

Tightness in chest

Difficulty breathing

Heart pounding

Nausea

Trembling

Symptoms of
fight or flight

Homeostasis

Most of the human body's organs and collective systems are working towards the same goal: to keep everything the same. This property of the body is called homeostasis, a word derived from Greek words that mean 'standing still the same'. Metabolism requires a relatively narrow set of conditions to work effectively and keep the body alive. Homeostasis uses the nervous and endocrine systems to maintain those conditions.

The breathing rate maintains blood oxygen levels, and the heart rate maintains a blood pressure that delivers oxygen to the body. Insulin and glucagon ensure that the amount of blood sugar stays close to a fixed level, while the pituitary's anti-diuretic hormone (ADH) keeps the level of the body's water content (and the concentration of the chemicals dissolved in it) relatively constant. Finally, like all mammals, humans are endothermic, meaning the body stays at a steady temperature, using its energy supply to heat and cool the body as required (see page 230).

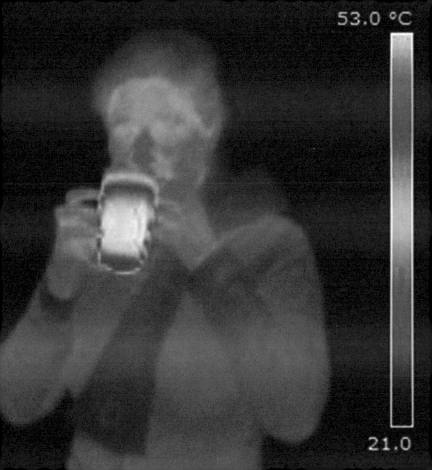

Negative feedback

Homeostasis largely works through a self-regulating system known as negative feedback. When a process creates an imbalance of some kind, the body is stimulated to promote an opposing process to correct that imbalance.

This process is not exclusive to the human body: the ballcock and valve in a toilet cistern make a simple negative-feedback loop. A low water level, as indicated by the ballcock, stimulates a high flow rate of water through the refill valve. As the water level rises, the rate of refilling slows until it stops altogether. There are multiple examples of feedback loops in the body. One of them concerns activity level. When activity needs to be raised to tackle a stressful situation, the hypothalamus and pituitary glands release a cascade of hormones to stimulate the thyroid gland. In addition, the adrenal gland is stimulated to make cortisol. Rising levels of cortisol inhibit the brain-based glands from producing their hormones, and activity levels revert to normal – unless, that is, the source of stress persists.

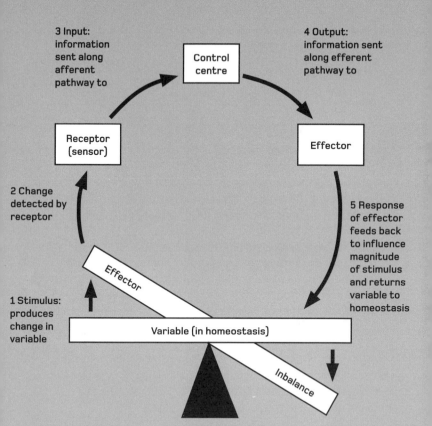

Breathing rate

Breathing is entirely unconscious, which is why you do it in your sleep. When the body is relaxed, it requires less oxygen, and so the breathing rate drops. However, when the body is working hard – especially the muscles – the body requires a lot more oxygen to burn fuel at a higher rate, and so the breathing rate (and volume) automatically increases.

The rate of breathing is controlled by the medulla oblongata in the brainstem. Chemical receptors respond to acidity of the blood. Carbon dioxide produced by cellular respiration dissolves in the blood, making it acidic. A boost in energy consumption increases this acidity, and the brain responds by increasing the breathing rate – primarily to rid itself of carbon dioxide, but increasing oxygen levels in doing so. Because it relies on carbon dioxide levels, the system does not respond well to abnormally high oxygen levels, such as from breathing pure oxygen, which would saturate the blood and lead to chemical damage in the lungs, brain and eyes.

Respiratory rate is based on the number of breaths a person takes per minute. The normal rate of respiration for an adult at rest is 12 to 20 breaths per minute, and all else being equal, there is a clear link between respiratory rate and heart rate.

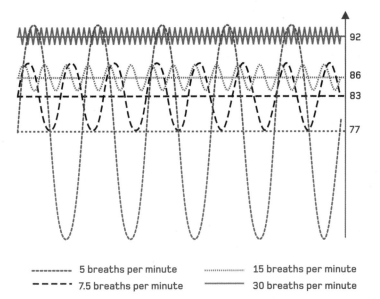

AVERAGE HEART RATE (beats per minute)

---------- 5 breaths per minute   ··············· 15 breaths per minute

— — — — 7.5 breaths per minute   ————— 30 breaths per minute

Temperature control

The human body functions best at 37°C (98.6°F). This is measured in the mouth – the core of the body is a little warmer. The biochemistry in our cells is fine-tuned to work at this temperature, the heat itself coming from the chemical activity of the cells. However, the air surrounding a body seldom matches this temperature and the body becomes either too hot or too cold as a result.

Temperature control, or thermoregulation, is managed by the hypothalamus (see page 156) using inputs from receptors in the skin. Too warm, and the body increases the rate of heat loss by dilating blood vessels under the skin to bring warm blood closer to the air. Sweating puts water onto the skin, which evaporates, taking energy with it. Too cold, and sweating is inhibited; the blood vessels in the skin constrict, keeping the blood away from the body periphery. Hairs on the skin stand erect to trap an insulating layer of still air around the body. In extremis, the skeletal muscles spasm, or shiver, releasing heat to warm the body core.

The hypothalamus regulates the body at a standard temperature of 36.5–37.5 °C (97.5–99.5 °F)

Hypothalamus

Cold peripheries trigger heat retention

Hot peripheries trigger heat loss

Skin surface capillaries constrict, conserving heat

As blood temperature approaches the point regulated by the hypothalamus, heat loss and retention mechanisms return to baseline state

Capillaries dilate, releasing heat to surroundings

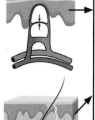

Sweat secretion ceases and hairs stand up, trapping a layer of insulating air

Sweat glands activate, causing heat loss by evaporation

Osmoregulation

Osmoregulation is the homeostasis of water concentration. As with other regulatory functions, this is controlled by the hypothalamus. This region of the brain uses osmoreceptors to check water levels in the blood by detecting osmotic pressure. This is the force by which water diffuses in or out of cells according to the saturation of chemicals dissolved in it.

When osmotic pressure is high, water levels are low, and the hypothalamus signals the pituitary gland to release anti-diuretic hormone (ADH). This constricts in the peripheral veins near the skin so less water is available for sweating. However, ADH's primary role is to stop the kidneys removing too much water from the blood to make urine. Instead, the kidneys put that water back into the blood supply. In a classic negative-feedback loop (see page 226), this results in an increase in the water content of the blood and body, and reduces the osmotic pressure. Therefore, the pituitary stops production of ADH and the kidneys begin to return to removing water from the body.

How osmoregulation works

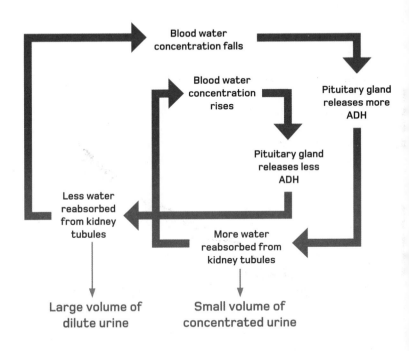

The kidneys

Excretion is the removal of waste products from the body. The main waste product is urea – a toxic, nitrogen-rich chemical produced when the liver processes proteins. Urea and other waste are removed by the kidneys, which are two bean-shaped structures located under the diaphragm either side of the spine. They filter unwanted materials from the blood and mix it with water to create urine, a liquid waste.

A kidney contains thousands of tubular nephrons. Each one begins at a glomerulus, a cup-shaped filter lined with capillaries. Water and other chemicals, including urea, leak out of the capillaries and enter a convoluted nephron. As the liquid moves through the tube, useful substances are reabsorbed and returned to the blood. At the long loop of Henlé water is reabsorbed (if needed), and the resulting mixture of urea, salts and water drips from the nephron into a collecting duct that serves many nephrons at once. The collecting ducts converge and empty into the ureter – a long tube taking urine from the kidney to the bladder – for ultimate removal.

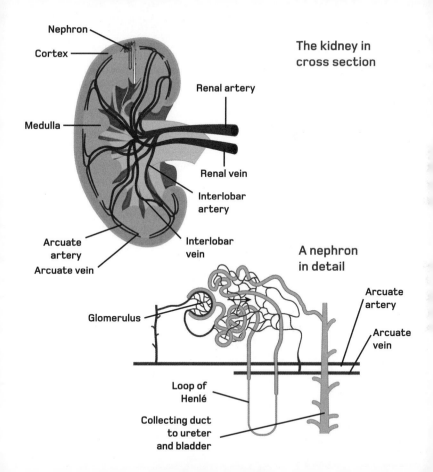

Nephron

Cortex

Medulla

Renal artery

Renal vein

Interlobar artery

Arcuate artery

Arcuate vein

Interlobar vein

The kidney in cross section

A nephron in detail

Glomerulus

Arcuate artery

Arcuate vein

Loop of Henlé

Collecting duct to ureter and bladder

The bladder

The bladder is a muscular bag with a volume of between 300 and 500 millilitres (10–18 fl oz). Its job is to store urine and then remove it from the body during urination. The bladder is located in the pelvis on the midline of the body. It receives urine from each kidney via long vessels called ureters, their muscles squeezing the urine into the bladder using peristalsis.

The ureters enter at the top of the bladder and the single exit point, the urethra, is centrally located at the bottom. The urethra is kept closed by a muscular sphincter. Once the bladder becomes half full, the weight on the sphincter triggers a nervous response that stimulates a desire to urinate. This grows in intensity as the bladder fills. The urethral sphincter is striated muscle that is under voluntary control. Opening the sphincter allows the urine to leave. It exits the body at the urinary meatus, which is just behind the clitoris in women and at the tip of the penis in men. In women, the bladder drains mostly under the effects of gravity, while men use muscles at the base of the penis.

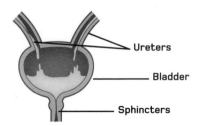

Ureters

Bladder

Sphincters

Under normal circumstances, the bladder fills steadily with urine from the kidneys.

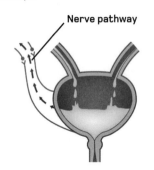

Nerve pathway

As the bladder fills and enlarges, nerve signals are sent to the brain.

The sphincters relax, and urine flows into the urethra, emptying the bladder.

Urethra

The reproductive system

While other body systems are concerned with the survival of the body, the reproductive system plays no role in that goal. Instead, it diverts energy towards another purpose: to produce children. The male and female reproductive systems are entirely different. During development, sex hormones – oestrogens in females and androgens, such as testosterone, in males – cause the same set of embryonic tissues to form homologous structures. Both sexes have two gonads, or sex glands (ovaries in female; testes in males). These produce gametes (sex cells) that combine to form new humans.

The male sex organs are entirely devoted to producing and distributing male sex cells, or gametes (sperm). By contrast, the female reproductive organs contain a uterus, or womb. This chamber sits between the ovaries and the vagina, and is where an embryo, and then fetus, will be supported and protected as it develops into a baby capable of living outside the mother. Once that is achieved, the baby will leave via the vagina.

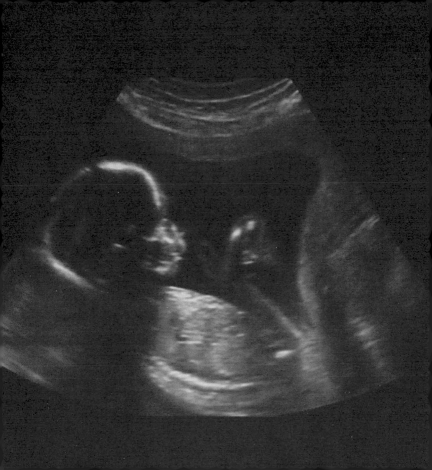

Female sex organs

The female sex organs are mostly contained within the body. The visible external feature is the vulva, which surrounds the opening to the vagina (and the urethra). The opening has two sets of vertical lip-shaped structures, known as the labia. At the top of the labia is the clitoris, the main erectile tissue in the female genitals. Physical contact makes the clitoris become engorged with blood and heightens arousal during sex.

The vagina is a muscular tract that leads to the cervix, the narrow opening to the uterus, or womb. The uterus is an elastic chamber in which a fetus develops during pregnancy. When a women is not pregnant, it is pear shaped and about 7 centimetres long and 4.5 centimetres across at the widest point at the top (2¾ x 1¾ in). The ovaries, which produce eggs and female hormones, are located either side of the uterus. They each connect to it via an oviduct, also known as a fallopian tube. These are funnel-shaped near the ovaries so they can collect any eggs produced, and provide a route to the uterus.

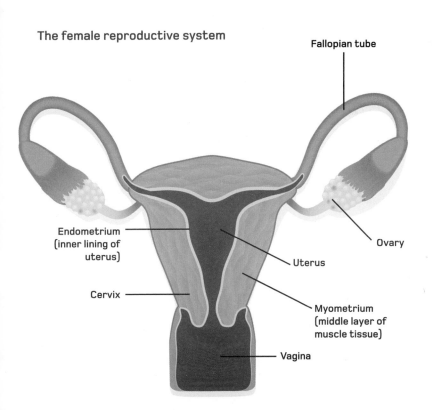

The female reproductive system

Fallopian tube

Ovary

Endometrium
(inner lining of
uterus)

Uterus

Cervix

Myometrium
(middle layer of
muscle tissue)

Vagina

Male sex organs

The male reproductive system contains the testes, epididymis, the prostate gland and the penis. The testes hang outside the body cavity in a sac called the scrotum.

The testes are endocrine glands that produce hormones for creating masculine characteristics elsewhere in the body. Their primary role is to produce sperm, of which they create 100 million every day. Sperm cells from each testis pass into a convoluted tube called the epididymis, through which they travel for about three months. The epididymis connects to the prostate gland via a tube called the vas deferens. During ejaculation, sperm are pumped up the vas deferens into the prostate, where they are mixed with a milky fluid supplied by the seminal vesicles, forming semen. This is pushed into the urethra and a series of muscle contractions force it out of the tip of the penis. During the sexual arousal that leads to ejaculation, the penis becomes erect and hard. This is the result of an increase in blood pressure within spongy erectile tissue surrounding the urethra.

The male reproductive system

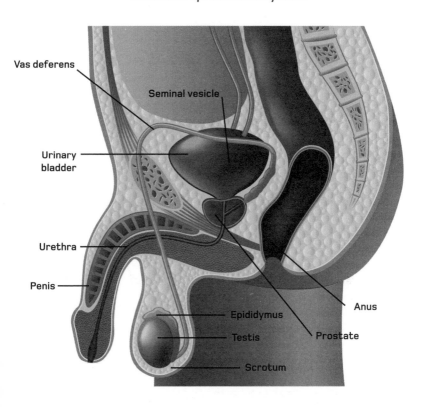

- Vas deferens
- Seminal vesicle
- Urinary bladder
- Urethra
- Penis
- Epididymus
- Testis
- Scrotum
- Anus
- Prostate

The human egg

The egg, also known as an ovum, is made by the ovary. As a sex cell, it contains half the amount of genetic material that other body cells have (23 chromosomes; see page 256). It is destined to meet a male sex cell and merge their genes to create a zygote, the first cell of a new individual.

While sperm are produced in huge quantities, human ovaries have about 400,000 follicles that can grow into ova, just one or two of which ripen for releases each menstrual cycle. In all, the average woman produces about 360 mature eggs in her lifetime. At 0.12 millimetres wide, the ovum is the largest single cell made by the human body, just about visible with the naked eye. The size of the egg shows what a big job it has to do. While a sperm carries chromosomes for the zygote, the ovum must do everything else to support it. The cell contains the nutrients and cellular equipment required to power the growth of a new individual. This material is stored in the cell's voluminous cytoplasm, specifically the ooplasm, and can be understood as the egg's yolk.

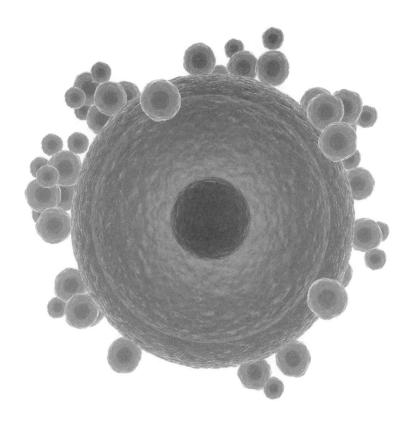

Ovulation

Ovulation is the process of releasing an ovum from the ovary. An ovum is a haploid cell, which means it contains half the chromosomes of other body cells. Haploid cells are created by a double cell division called meiosis. Meiosis can produce four cells; two are from the first division and they both then divide again. However, in human egg production, only one cell is produced. At each division, one of the cells hoards the cytoplasm, and the other cell is just a tiny packet of genetic material called a polar body. This system allows the egg to reach its great size.

The division process starts before birth, where primary oocytes are formed, but then stops until puberty. The hormones involved in the menstrual cycle then stimulate one of the oocytes to continue the divisions. As it does so, a follicle develops around it, which eventually merges with the surface of the ovary. The egg can leave the ovary and enter the oviduct while completing the last stage of division. The follicle also escapes, forming a corpus luteum (see page 248), to aid any zygote that may be formed.

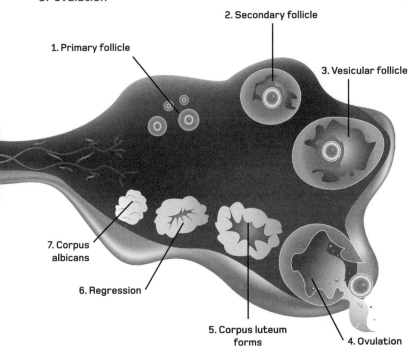

Stages
of ovulation

1. Primary follicle

2. Secondary follicle

3. Vesicular follicle

4. Ovulation

5. Corpus luteum
forms

6. Regression

7. Corpus
albicans

The menstrual cycle

The female reproductive system cannot produce a baby at any time. The lining of the uterus needs to be prepared to receive an embryo so that it may complete its development. As a result, the female reproductive system goes through a complex cycle, lasting on average 28 days. The cycle, called a period, is mediated by four hormones: oestrogen, progesterone, follicle stimulating hormone (FSH) and luteinizing hormone (LH).

The cycle begins with the development of a follicle in the ovary, stimulated by low levels of FSH, LH and oestrogen. This starts to thicken the lining of the uterus (endometrium), making it highly vascularized. Around day 14, a surge in the three hormones results in ovulation. The ovum is viable for the next 20 hours, while the corpus luteum produces progesterone to inhibit the other hormones. If no sperm fertilizes the ovum, the corpus luteum degrades after 10 days, forming the scar-like corpus albicans. The fall in progesterone causes the endometrium to be shed through the vagina, resulting in a week of bleeding, known as menstruation.

Day in cycle 1 2 3 4 5 6 7 8 9 10 11 12 13 14 15 16 17 18 19 20 21 22 23 24 25 26 27 28

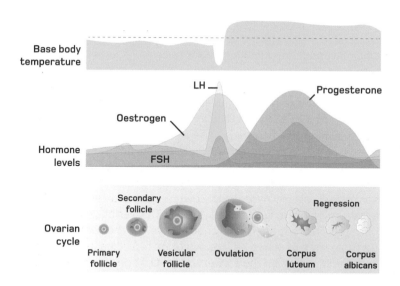

Sperm

Spermatazoa, or sperm for short, are the male gametes, or sex cells. Their single purpose is to carry a half-set of chromosomes to the egg. It takes about 70 days for a sperm to develop inside the testes, with a single starting cell – the spermatogonium – producing four sperm.

Human sperm are 0.05 millimetres long, considerably smaller than the ovum, especially considering most of this length is made up of a flagellum. This is a whip-like tail that propels the cell through the mucus lining the uterus as it heads in search of an ovum. Not all sperm swim in the right direction, but those that do travel up to 4 millimetres a minute, although many are slower. The journey through the uterus is 17.5 centimetres (7 in), so the fastest sperm reach the ovum in less than an hour; others arrive anything up to three days later. This variation has a purpose. The fast sperm may not always win the race – they may be too early and arrive before ovulation has released the egg. In the end, it may be the slower sperm that get there on time.

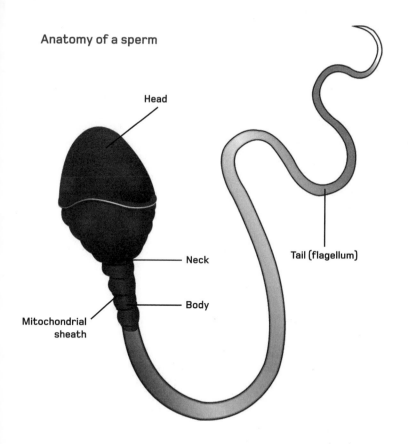

Anatomy of a sperm

Head

Neck

Body

Mitochondrial
sheath

Tail (flagellum)

Copulation

For sperm and egg to meet, sperm must be delivered to the cervix, the opening between the vagina and the uterus. This opening softens and widens as ovulation approaches, to allow for the passage of sperm. The process of getting sperm to the cervix is called copulation or sexual intercourse. It begins with sexual arousal. In men, this involves the penis becoming hard and erect; in women, the vulva – the opening of the vagina – swells and the vagina lengthens, making it easier for the penis to be inserted.

The bare minimum that is required next is for the man to achieve orgasm. This is done by rhythmically rubbing the penis in and out of the vagina, resulting in an increasingly pleasurable sensation that climaxes with the ejaculation of semen. The female orgasm is achieved by a rhythmic rubbing of the clitoris (an erectile tissue at the top of the vulva) and the stretching of the vagina by the penis. The orgasm results in equally pleasurable sensations and pulsating contractions of the vagina and cervix. It is likely that these contractions help sperm enter into the uterus.

Sexual intercourse

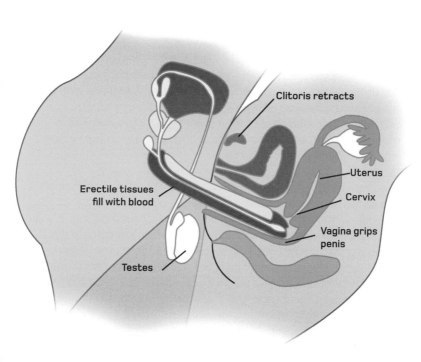

Clitoris retracts

Uterus

Cervix

Vagina grips penis

Erectile tissues fill with blood

Testes

Conception and implantation

Copulation delivers millions of sperm into the vagina, but only a few are successful in making the journey through the uterus and the correct oviduct to the egg. And only one will penetrate the outer layers of the egg to deliver its genetic cargo. Once it is inside, all other sperm are locked out. This is the moment of conception, and the egg has been transformed into a zygote, the first cell of a new human body.

The zygote begins to grow 24 hours later, by which time it has moved halfway along the oviduct. It divides again every eight hours and, upon entering the uterus three days after fertilization, is a ball of 16 cells called a morula. Meanwhile, hormones from the corpus luteum maintain the right conditions in the uterus. By day four, the enlarged ball of cells is a blastocyst, an inner mass forming where the embryo will ultimately develop. The blastocyst sheds an outer membrane inherited from the ovum. On day six it settles on the endometrium and over the next week it becomes implanted so it can receive oxygen and nutrients from the mother.

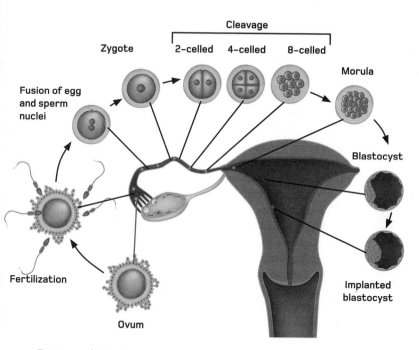

Cleavage

Zygote · 2-celled · 4-celled · 8-celled

Morula

Fusion of egg and sperm nuclei

Blastocyst

Fertilization

Ovum

Implanted blastocyst

From ovulation to implantation

Cell division

The essential driver of human growth and development is a kind of cell division called mitosis. The process involves several complex stages and ends with two daughter cells, each of which contains an identical copy of the genetic material of the parent cell.

As the body grows, its cells specialize, with daughter cells having a specific purpose – being bone cells or blood cells, for example. In general, this process is controlled by parents that are stem cells – that is, unspecialized cells that are capable of dividing into a range of specialized cells. Once specialized, a cell can only divide into daughters with the same function.

To make sex cells, the body uses a different kind of cell division, known as meiosis. Here the cell divides into four, with each daughter containing just half of the genetic material of the parent. When two sex cells fuse during fertilization, they produce a single zygote cell, which has a full set of genes.

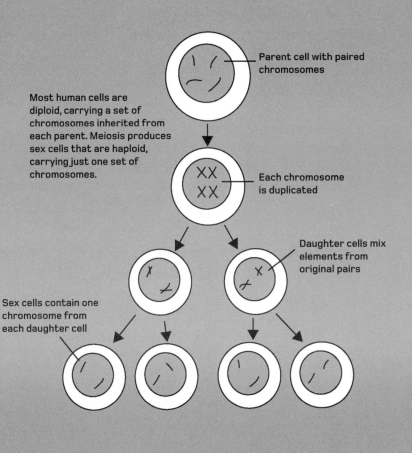

Parent cell with paired chromosomes

Most human cells are diploid, carrying a set of chromosomes inherited from each parent. Meiosis produces sex cells that are haploid, carrying just one set of chromosomes.

Each chromosome is duplicated

Daughter cells mix elements from original pairs

Sex cells contain one chromosome from each daughter cell

Contraction

Human beings use sexual intercourse for reasons other than reproduction. People have sex for pleasure; it is also a significant means of strengthening and maintaining the bond between partners. It is often the case, therefore, that sexual partners are not intending to achieve pregnancy every time they have intercourse. In addition, sexual contact can spread diseases of varying severity, such as HIV, herpes or chlamydia.

Contraception is the term for a variety of methods that prevent conception, some of which prevent diseases. There are many kinds, each suited to different circumstances. The safest is the condom, a latex sheath that fits over the penis to create a physical barrier to both sperm and disease. A diaphragm coated in spermicide fitted over the cervix does something similar, but offers less protection against disease. Contraceptive pills take over the hormonal control of the uterus, meaning embryos cannot implant there, while a surgical vasectomy disconnects the testes from the prostate so no sperm reach the semen.

Popular forms of
contraception

Oral contraception

Intrauterine
contraceptive
device

Calendar rhythm
method

Contraceptive patch

Surgical sterilization

Contraceptive
injection

Diaphragm

Condom

Female
condom

The placenta

At the start of the third week of pregnancy, the inner region of the blastocyst begins to differentiate into layers that form the internal and external tissues of the fetus. Meanwhile the outer region, known as the trophoblast, begins spreading branchlike into the endometrium. This is the early stage of a temporary organ called the placenta, which takes over from the corpus luteum in producing the progesterone that suspends the menstrual cycle during pregnancy. Fully grown, it weighs 500 grams (1 lb) and is 20 centimetres (8 in) long.

An umbilical cord connects the placenta to the fetus's navel. The cord contains two fetal arteries and one vein. Within the placenta, these fetal blood vessels extend into multiple column-shaped villi, surrounded by vessels supplied by the mother. This enables oxygen, nutrients, waste products and even hormones to pass between mother and fetus. Nutrients travel via the umbilical cord to the fetal liver, while oxygenated blood goes straight to the heart, bypassing the lungs, which do not function until birth.

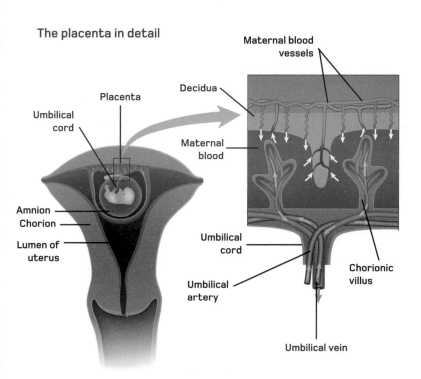

The placenta in detail

Placenta

Umbilical cord

Amnion

Chorion

Lumen of uterus

Maternal blood vessels

Decidua

Maternal blood

Umbilical cord

Umbilical artery

Chorionic villus

Umbilical vein

Ectopic pregnancy

An ectopic pregnancy is one in which an embryo implants outside the uterus, and the placenta develops in a place that cannot accommodate it. This kind of pregnancy always requires medical attention. It causes pain and unexpected vaginal bleeding, and a body-wide shock reaction. If left untreated it will damage the reproductive system and, in the great majority of cases, the fetus cannot survive.

Nearly all ectopic pregnancies occur in the oviduct, also known as a fallopian tube. Mostly, implantation takes place far too early, in the opening of the oviduct next to the ovary, and less often further down. The developing placenta blocks the oviduct. Embryos can also implant on the ovary itself, and so develop in the body cavity. Rarer ectopic pregnancies block the cervix or implant too deeply into the endometrium. Ectopic pregnancy appears to be linked to inflammation of the reproductive system, which can be caused by smoking and chlamydia (a sexually transmitted disease).

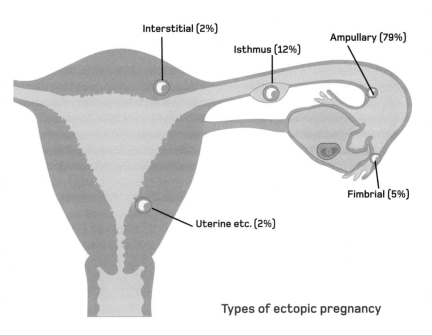

Interstitial (2%)

Isthmus (12%)

Ampullary (79%)

Fimbrial (5%)

Uterine etc. (2%)

Types of ectopic pregnancy

First trimester

A pregnancy's length is measured from the last menstrual period. However, conception actually occurs about 14 days after this start. A human pregnancy lasts 9 months, or 40 weeks. It is generally divided into three 3-month periods, known as trimesters. During the first trimester an embryo grows into a fetus that is between 7 and 10 centimetres ($2\frac{3}{4}$–4 in) long. There is no distinct transitional point between embryo and fetus, but once the developing baby becomes recognizably human, at about week 11, it is known as a fetus.

After the first month, the embryo is about the size of a grain of rice. It has developed a distinct head, with dark circles for eyes and a jaw. In the second month, the circulation system starts working. Limbs and digits begin to develop. Also in this month, the beginnings of all major body systems, including the brain and nervous system, digestive tracts and bones, appear. In the third month, the fetal body becomes fully formed, with all its features in place.

Months 1 to 3

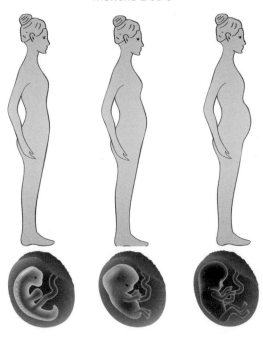

All the baby's major systems and structures develop during the first trimester, including the brain, spinal cord and heart. Arms and legs grow and facial features become distinct.

Second trimester

During the second trimester of a pregnancy the fetus grows to a length of around 30 centimetres (12 in). The body systems laid down in the first trimester become more defined with finer anatomical features. In month four, hair begins to grow, including eyelashes and eyebrows, and nails are forming on the fingers and toes. The baby can move, although it is still too small (15 cm/6 in long) for the mother to feel it.

In month five, the skin is covered in a fine hair called lanugo. These hairs fall away soon after birth. The muscles are now developing and the growing baby stretches and flexes them, making movements that the mother can detect.

In month six, the fetus has fingerprints, can open and close its eyes and even hiccups as the nervous system rehearses the motions of the diaphragm. If the baby is born prematurely during week 23, halfway through this month, it can survive if given intensive medical care.

Months 4 to 6

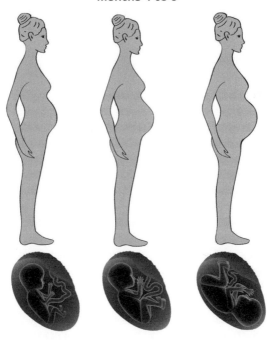

By the end of the second trimester the baby can hear and swallow. Nails appear on the fingers and toes and the baby has genitals. Muscle tissue and bones keep developing, and bones become harder.

Third trimester

The final trimester of a pregnancy is all about growth. The baby lengthens to an average of 48 centimetres (19 in), although it is tightly curled into a ball with legs raised to the abdomen. As well as increasing in length, the baby becomes considerably fatter, eventually achieving a weight in the region of 3.2 kilograms (7 lb). This fat will insulate the baby after birth, once it leaves the warm conditions of the uterus.

In month seven, the fetus becomes fully sensitive to external stimuli. It hears well through the uterine wall, responds to light and can feel pain. Babies born during the eighth month, after week 36, are described as pre-term, not premature, and have a very good chance of survival. All the internal organs are ready except the lungs, which may still be in development. During the ninth and final month, the baby repositions to prepare for birth. Having been upright through the pregnancy, it turns upside down, and descends so its head presses against the cervix. This weight helps to stimulate the conditions for birth.

Months 7 to 9

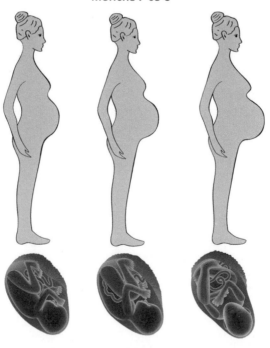

The baby's nervous system develops enough to control some body functions and the respiratory system reaches maturity. The heart and blood vessels are complete. Muscles and bones are fully developed.

Birth

The process that leads to birth is called labour. In the final weeks of pregnancy, the mother's pelvic girdle will have separated at the front to create more room. Labour generally begins gradually with small, infrequent uterine contractions.

The initial stage of labour develops slowly, with contractions becoming more frequent and intense over the course of several hours. This gives the cervix time to soften and dilate. Once it is fully dilated, around 10 centimetres (4 in), the women enters a transitional phase, a pause that gives her time to prepare for the second stage of labour, the birth itself. Painful contractions lasting 60 seconds or more occur every few minutes. They push the baby through the cervix and along the birth canal. Up to this point, the newborn baby has only received oxygen via the mother's blood. In the first few seconds of independent life the baby takes a breath, and its blood becomes oxygenated by its own lungs. The umbilical cord is cut, and minutes later the placenta detaches and leaves via the birth canal.

The three stages of childbirth

Stage 1: Cervical dilation

Waters break

Stage 2: Expulsion

**Stage 3:
Delivery of the placenta**

Growth

Growth is controlled by the growth hormone, which is secreted by the pituitary gland. It is an anabolic peptide that promotes the lengthening and hardening of bones, building of muscle mass, and ensures the liver maintains a supply of sugar to power all these changes. The exact rate of growth depends on the availability of raw materials – fats, proteins and carbohydrates – in the diet, and may be slowed by digestive diseases or a simple lack of food.

The fastest growth rate is in the first year, in which a baby will almost double in height and triple in weight. After that, growth slows to about 6 centimetres (2½ in) a year, with the child reaching half its adult height around the age of 2 and half adult weight at about 10. Growth is not constant, but appears in spurts. Both sexes are roughly the same size until the onset of puberty. During puberty, growth accelerates again, beginning in girls from about the age of 10 and 12 in boys. Girls reach full adult height at 15 or so, while boys continue growing until the age of 18.

Average height (m)

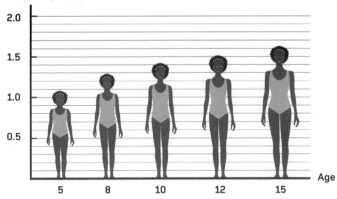

Average height (m)

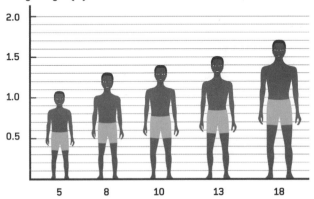

Female adolescence

Adolescence, or puberty, is the period in which the human body changes from the largely gender-neutral child form into the sexually mature adult form. It is also the time when young people learn to behave in adult ways and to experiment with taking control of their lives.

In girls, this generally starts around the age of 10 or 11 – although in past generations it was normal for it to occur a year or two later. One of the first sexual characteristics to appear is pubic hair (armpit hair develops a little later). The region under the nipples swells and the breasts begin to develop gradually. They can appear to be different sizes to begin with, but nearly always even out. The body also develops a higher percentage of fat. This fat increase is linked to the menarche, the onset of menstruation, which generally begins around the age of 13. The menarche does not indicate that the girl has started ovulating. Most girls produce eggs on a regular basis from the age of about 16.

Changes in girls during puberty

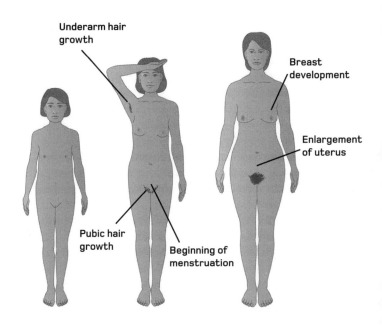

Male adolescence

Males generally begin the physical and emotional changes associated with puberty later than females, at around the age of 12 or 13. One of the first changes is an enlargement of the scrotum, which loosens so it hangs lower and often becomes darker in colour. The first ejaculation occurs around a year later, and the semen carries fertile sperm from the start. In later puberty the penis will lengthen and then widen.

The scrotum and pubic area develop short, coarse hairs. A couple of years later, hairs develop in the armpits, the hairs on the chest and forearms thicken and the first signs of a beard and moustache appear. The voice deepens.

A boy's growth during puberty results in him gaining height due to lengthening bones, but it takes longer for the muscles and nerves to develop around them. Boys are often gangly and ill-coordinated compared to girls until the middle of puberty, when increasing muscle mass makes them physically stronger.

Changes in boys during puberty

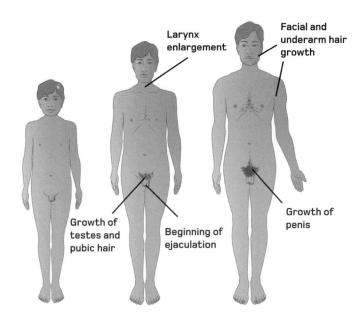

Larynx enlargement

Facial and underarm hair growth

Growth of testes and pubic hair

Beginning of ejaculation

Growth of penis

Mammary glands

The breasts, or mammary glands, are exocrine glands that produce milk to feed newborn babies. They develop during puberty but only produce milk, or lactate, following the birth of a child. Pregnancy hormones will have prepared the breasts for lactation. They grow larger as fat is laid down, and the ducts that carry milk from the alveoli (cell-lined cavities) deeper inside to the nipple widen into reservoirs, or sacs.

One of the hormones involved is prolactin. After birth, the other pregnancy hormones drop, but prolactin remains high, and this stimulates cells around the mammary alveoli to secrete milk. Lactation is further stimulated by the baby sucking on the nipple, which releases milk through a series of tiny pores. Human breast milk is an emulsion of water and about 14 per cent fat taken from the surrounding breast. The milk also contains many other nutrients and antibodies from the mother's immune system. These are especially high in the first milk, or colostrum.

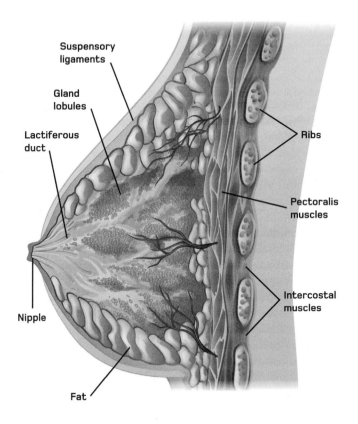

Suspensory ligaments

Gland lobules

Lactiferous duct

Nipple

Fat

Ribs

Pectoralis muscles

Intercostal muscles

Sexuality

Besides the physical aspects of puberty, adolescence also involves the awakening of a person's sexuality, where he or she becomes attracted to other people and seeks to engage in sexual behaviour with them.

Engaging in sex acts with a member of the opposite sex is described as heterosexual behaviour, while having sex with a member of the same sex is homosexual (*hetero* means 'other'; *homo* means 'same'). People described as bisexual are attracted to both sexes. Sexual orientation is seldom fixed to begin with; young people report attraction to the same sex and to the opposite sex at different times. The roots of sexuality are far from understood, but sexual preference is today generally thought to have deep-seated developmental causes. Evidence for a genetic contribution remains uncertain, but studies have reported different brain structures in homosexual men and women that suggest sexuality is certainly determined in early fetal development, perhaps by hormone levels in the uterus.

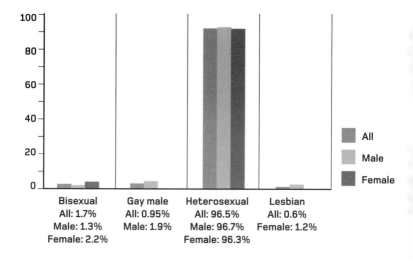

Sexual orientation by gender*

100
80
60
40
20
0

Bisexual
All: 1.7%
Male: 1.3%
Female: 2.2%

Gay male
All: 0.95%
Male: 1.9%

Heterosexual
All: 96.5%
Male: 96.7%
Female: 96.3%

Lesbian
All: 0.6%
Female: 1.2%

All
Male
Female

* Based on a 2012–13 survey of
20,055 Australians aged 16–69

Sex and gender

The concept of gender is not the same as the idea of sex. A person's sex is defined as male or female by the presence of one or other set of genitals. Gender, however, is *traditionally* divided into masculine and feminine, and is defined by a complex set of conventions and behaviours that often tend to be cultural rather than biological. Mode of dress and hairstyle are two obvious examples.

However, it is estimated that one to two per cent of humans are intersex – that is, they have some combination of both male and female sex organs and exhibit a mix of male and female secondary sexual traits such as typically male or female hip and shoulder shape, facial and body hair and breast development. A smaller group of 'gender dysphoric' people feel their physical body is the wrong sex for their gender identity. Such people may take the radical step of altering their body's sexual characteristics, either surgically or medically. Others opt to identify with a third gender that is neither masculine nor feminine.

Aspects of gender and sexuality

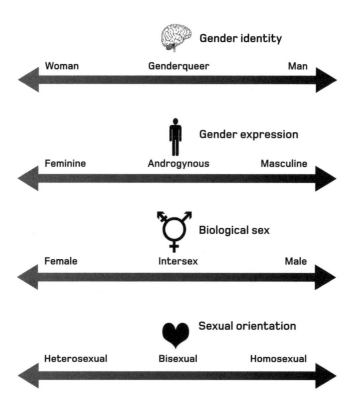

Inheritance

One can tell simply by looking that humans inherit the characteristics of their parents. Each child is unique, but contains a blend of features – hair colour, eye colour, height, build – shared by their parents and earlier ancestors. The mechanisms by which these features are passed from one generation to the next is investigated by the science of genetics.

Geneticists study genes, a set of coded instructions that guide the growth and development of all living things. However, the concept of the gene is more slippery than that. Genetic code is held in an intricate molecule called DNA (see page 286). A gene is the unit of inheritance, and so one gene could be seen as a specific strand of this chemical. It could also be a measurable characteristic that passes from parent to child. The true focus of genetics is to identify how strands of DNA end up creating body features. To that end, geneticists have decoded the entire human genome – the complete set of genes in a human body – an estimated 20,000 genes.

DNA

DNA is short for deoxyribonucleic acid, the chemical within which genetic code is stored. There are 2 metres (6½ ft) of it in each human cell. Added together, the total DNA in your body would stretch to the Sun and back dozens of times over.

The great majority of a cell's DNA is housed in the nucleus, with a tiny amount in the mitochondria. DNA was first isolated in 1869 in the pus of infected wounds. It took another 80 years to reveal that it had a double-helix structure. The sides of this helix are chains of ribose sugar bonded together with phosphates. The crossbeams of the spiral are made of pairs of four nucleic acids: adenine pairs with thymine; cytosine always links to guanine. These four 'bases' make a simple four-letter code using ATCG. The pattern of these chemical letters running through the DNA molecule is what creates a genetic code. The cell deciphers the code of each gene and uses it to construct a specific protein molecule – that is, a molecule with a specific metabolic job to do.

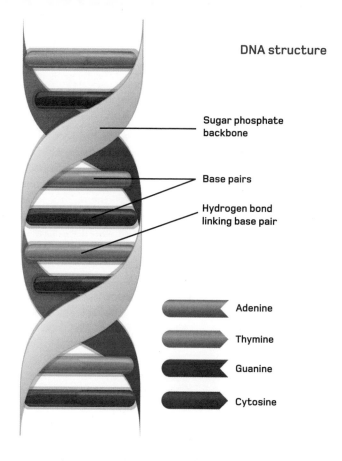

DNA structure

Sugar phosphate backbone

Base pairs

Hydrogen bond linking base pair

Adenine

Thymine

Guanine

Cytosine

Genes

The way most people understand a gene is through the expression of a measurable characteristic, such as hair colour. A child with the same hair colour as the parent has inherited that gene from them. This loose definition is used to track inheritable characteristics through a family, but it does not tackle the mechanism by which a piece of DNA that has moved from one generation to the next results in a specific characteristic.

When we look at the way the coded information held on DNA is read by a cell, we find that it is seldom the case that one piece of DNA results in a specific characteristic. Instead, a gene in this context relates to a strand of DNA that carries the information needed to make one protein, or more likely, one section of protein, known as a peptide. The four-letter ACTG code is divided into three-letter units, each one relating to a type of amino acid – the building blocks of proteins. A string of DNA 'letters' is therefore translated into a chain of amino acids in a specific order that forms a particular protein unit.

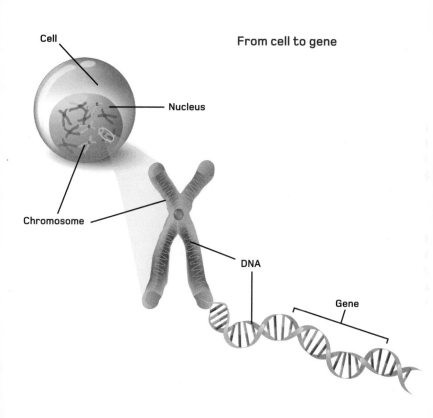

From cell to gene

Cell

Nucleus

Chromosome

DNA

Gene

Chromosomes

DNA is delicate and must be handled with care. Human DNA, in common with that of all multicellular life, is stored on scaffold-like structures called chromosomes. Every human cell has 46 chromosomes, 23 inherited from each parent. One half set of 23 chromosomes carries a complete set of genes, so every cell has a double set of genes (see page 292).

Chromosomes got their name from early cell researchers, who saw dark-coloured objects appear in a nucleus during cell division; *chroma* means 'colour' in Greek. Most of the time, individual chromosomes are too narrow to see, existing as a nebulous mass of chromatin, but this thickens prior to cell division.

The ultrafine DNA molecules are coiled around spindle-like proteins called histones. Clusters of histones form larger units called nucleosomes and strings of nucleosomes are themselves coiled – and then coiled again. This so-called supercoil is how such as enormous length of DNA can fit into the nucleus of a cell.

The nucleus of each diploid human cell contains 22 pairs of autosomal (non-sex) chromosomes and one pair of sex chromosomes.

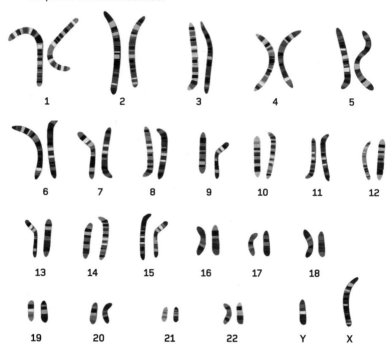

Genotype

Genotype is a way of understanding which genes have been passed from one generation to another. Every human body has received two sets of genes, one from its mother and another from its father. Different versions of the same gene – for eye colour, for example – are called alleles. Together, two alleles make up the genotype for any given gene.

A mother's alleles are not always the same as a father's. In rare cases, the alleles are co-dominant, and their influence carries equal weight. More often, however, one allele is dominant over the other, which is recessive. For example, the genotype for the eye colour gene may be Bb, with a brown allele (B) coming from the mother, and a blue one (b) from the father. B is dominant over b, so that genotype leads to brown eyes. When the time comes for those genes to be passed on, only one allele from the genotype will be added to a sperm or egg, entirely at random. This could result in a child with blue eyes, if the subsequent genotype is made of two recessive (b) alleles.

Inheritance of eye colour

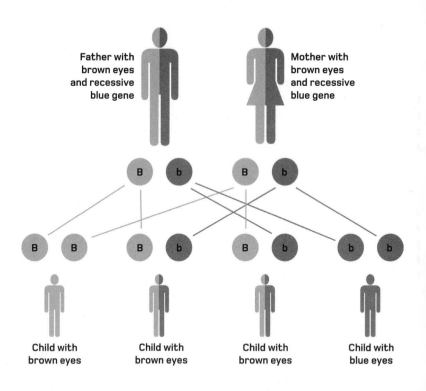

Father with brown eyes and recessive blue gene

Mother with brown eyes and recessive blue gene

B   b   B   b

B   B   B   b   B   b   b   b

Child with brown eyes

Child with brown eyes

Child with brown eyes

Child with blue eyes

Phenotype

The phenotype concerns a different way of looking at inheritance. While a genotype tracks the movements of DNA through the generations, a phenotype is concerned with the outward, physical manifestations of those genes. A phenotype is always the product of a genotype, but the process by which the latter leads to the former is generally very complicated to unravel.

A human phenotype includes characteristics such as the colour of hair, eyes and skin, height, and even intelligence. All these traits can be shown to pass through families. While some of them have been linked to specific isolated strands of DNA, for other characteristics the link remains unknown. There must be one, but it may involve the interaction of several genes at once – working together as a super-gene.

One area where the phenotype is especially important is in inherited diseases. Understanding how faulty genes cause these diseases can help in developing therapies for them.

Sex determination

Sex is determined by a pair of chromosomes labelled X and Y. A female has two X chromosomes; a male has an X and a Y. Females are homogametic, meaning all her eggs will contain an X chromosome, so every child gets at least one X. The male is heterogametic in that half his sperm carry an X and the other half hold a Y. Therefore, there is a 50–50 chance of inheriting an X and being female or getting a Y and becoming a male.

While a father's 22 other chromosomes match the 22 sent by the mother, gene for gene, his Y chromosome is much smaller than her X. The X carries many functioning genes, but the function of the Y's genes is to alter the development of the body from the female form into the male form. In female cells, the two Xs create a normal genotype with two alleles, but the Y does not carry a matching allele in male cells. Inherited disorders, such as colour-blindness and haemophilia, which are carried on the X chromosome are more common in men, because there is no alternative allele present to block them from developing.

Blood types

Every person has one of four blood types: A, AB, B or O. They are an inherited trait, and the genes involved are an object lesson in genetics. The types relate to antigens, or marker chemicals, that appear on the surface of red blood cells.

There are also antibodies that roam the bloodstream in search of aliens – things that have different antigens to the blood. So a person with A blood has A antigens on their cells and B antibodies in the bloodstream. These antibodies lock onto any cells with the B antigen, alerting the immune system. B-type blood does the exact opposite. Blood types are controlled by the genes A, B and O. The genotypes AA or AO result in the A blood type; B blood is from the BB or BO genotype. Inheriting an A and a B allele produces AB blood, with A and B antigens and no antibodies. The genotype OO results in O blood type, where the cells have no antigens, and the blood contains both antibodies. This means that O blood can be transfused safely into anyone, while people with AB blood accept all other blood types.

Blood groups

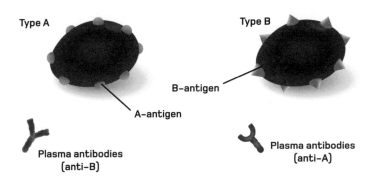

Type A
A-antigen
Plasma antibodies
(anti-B)

Type B
B-antigen
Plasma antibodies
(anti-A)

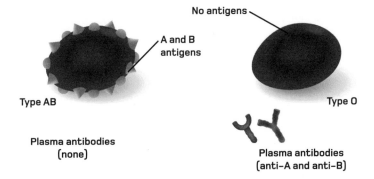

Type AB
A and B
antigens
Plasma antibodies
(none)

No antigens
Type O
Plasma antibodies
(anti-A and anti-B)

Nature vs nurture

The influence of genes and upbringing is the subject of a perennial debate. Which is the more powerful, the gene (nature) or the developmental environment (nurture)? The answer is seldom satisfying – it's always a bit of both. However, for certain characteristics, the influence of the environment is generally secondary. For example, the genes that control body symmetry and the arrangement of limbs and fingers are barely influenced by nature – although some chemicals can have a devastating effect here. But for other traits is it the other way around – things like personality, musical ability and temperament?

This second group of traits are usually the subject in a nature versus nature discussion, and as yet there is no strong evidence that they are genetic. The evidence is muddied, however, by the nature of the brain: this incredibly plastic organ can alter its own structure throughout life, making it hard to pin down whether the brain structure of a musically minded, extrovert hothead is down to genetics or other influences.

Human behaviour

While the major organs of the human body and other body systems are not particularly different to those of our mammal cousins, we humans see ourselves apart from the animals, as a naked, upright ape that has transcended biology.

While other life forms are dependent on their habitat for survival, human beings simply make themselves one in which to live. In so doing, having established a permanent home in Antarctica in the 1950s, the human is the only animal to live on all seven continents of Earth.

The keys to human success are bipedality – we can walk a long way while carrying and manipulating objects – and the ability to communicate ideas. It is at least possible that other animals can think. However, they do not appear to be able to communicate to each other as effectively as humans can. With a look, a gesture or a few words, one human can signal the content of his or her mind to others and, in so doing, conquer the world.

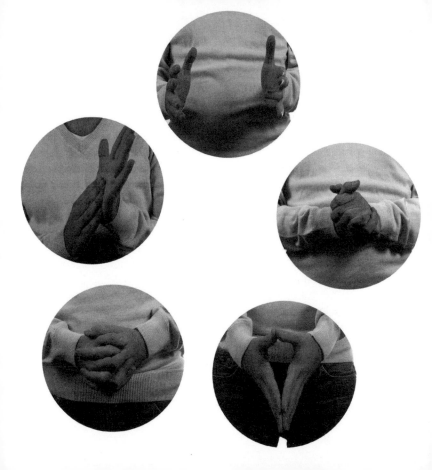

Social groups

The human social group arose from the need to defend resources. A single human would struggle to find enough food, water and shelter alone, and if he or she did, then a group of humans could simply take it away. In this respect, the human group might resemble a wolf pack, where animals work as a team. However, the size of a human community outstrips that of a pack. One reason for this is that it takes a lot longer for human young to mature, so a larger team is needed to ensure survival.

Within larger groups, smaller families or clan groups arose. Clans joined together for some roles but stayed separate for others, such as foraging and the raising of young. Certain baboon species that live in arid savannah adopt similar social structures. As the number and size of human groups grew, the scope for damaging conflicts within and between them also rose, and was countered by a developing tolerance of outsiders. Unseen in even our nearest relatives, this trait has served us well, allowing the formation of complex webs of social interactions that typify modern society.

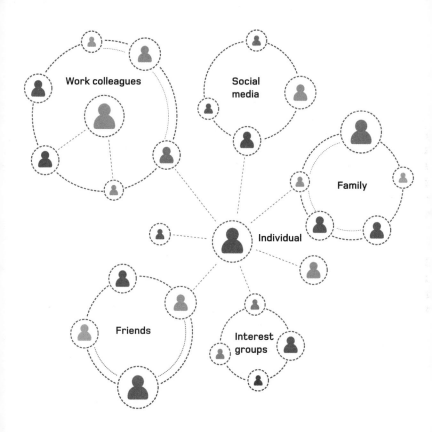

Language

There are several thousand human languages but no non-human ones. Some animals communicate with 'word' sounds, but none create sentences. Language is subject to the 'translation paradox', where words have a meaning that can only be defined using other words. Therefore, language must have arisen initially from a tacit agreement as to what it all means.

The evolution of language required a particular pharynx structure and fine control over the breath to make the voice clear. This appears to have developed after our species diverged from the last common ancestor with chimpanzees. More complex is the role of the brain in speech. Broca's area, a region of the frontal lobe, is involved in the physical production of speech. Wernicke's area, in the temporal lobe, concerns the association of words in producing and comprehending speech. Brain scans show that these areas are also associated with the mental processes of toolmaking, such as knapping a flint. This could suggest that language evolved as a way of teaching tool technology.

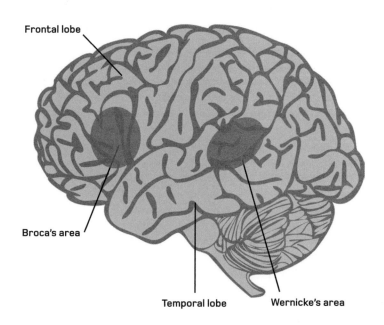

Language areas of the brain

Frontal lobe

Broca's area

Temporal lobe

Wernicke's area

The vocal cords

Humans talk using the vocal folds, better known as cords, inside the larynx. The larynx, sometimes seen on the throat as the Adam's apple, is a section of the windpipe, just above the epiglottis. The vocal cords are flaps of cartilage that can move in and out of the airflow leaving the lungs.

As the flaps close over the airflow, pressure builds up beneath. Eventually, the air breaks through the gap between the folds, making them vibrate. This vibration is passed to the air leaving the throat, creating an audible tone. The tone is then amplified by the throat, mouth and sinuses. Your ears hear your voice as it is still passing through these areas, and it has become slightly altered by the time it leaves your mouth. That is why your voice sounds different to you when recorded and played back. The tongue, lips, teeth and jaw all play a part in modifying the sounds emerging from the vocal cords, creating the phonemes – or letter sounds – that are used to make words.

Structure of the larynx

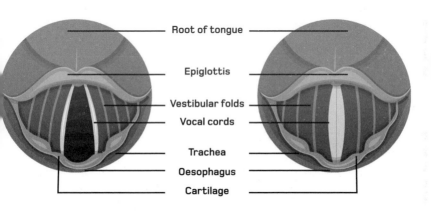

Root of tongue

Epiglottis

Vestibular folds

Vocal cords

Trachea

Oesophagus

Cartilage

When the vocal cords are open, air pases through the folds and causs them to vibrate.

When the cords are closed, lack of vibration means that air leaves the windpipe quietly.

Circadian rhythm

The human body has an internal clock located in the suprachiasmatic nucleus (SCN), which is part of the hypothalamus. This clock is used to control the body's cycle of sleep and wakefulness via a hormone called melatonin, which is secreted by the nearby pineal body. Melatonin levels rise in the evening, reducing blood pressure and making us feel sleepy. They drop again as dawn approaches, boosting blood pressure as we wake up. As melatonin decreases during the night, levels of cortisol (a hormone that prepares the body for activity) rise.

This clock maintains a steady rhythm that varies from person to person, but is close to 24 hours and is therefore circadian, meaning 'around a day'. The SCN is constantly being synchronized with actual light levels, with inputs from the eyes arriving via the optic chiasma located above it in the brain. Jet lag results from a major mismatch between light levels and the SCN's clock. Exposure to bright light is a good way to get the system in sync again.

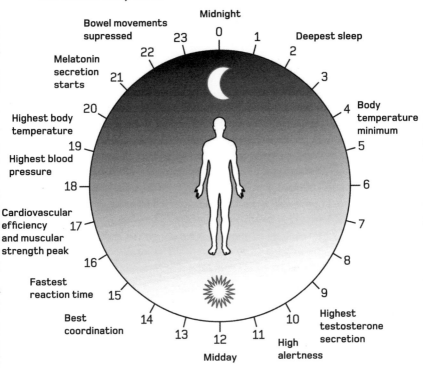

The human body clock

Sleeping

Humans are diurnal animals, and as night falls, we become sleepy. Falling asleep results in an altered state of arousal, in which we become unconscious of our surroundings. Sensory systems and voluntary movements do not shut down completely, but are highly reduced or inhibited. Heart rate, blood pressure and breathing rate slow to levels below those seen during wakefulness. The brain begins to behave differently. The alpha waves of resting wakefulness and beta waves of active thought are replaced with the lower-frequency theta waves associated with drowsiness. Nevertheless, sleep is an active process, in which the brain and the body cycle through a set of states that include periods of deep relaxation and dreaming.

The amount of sleep a person needs varies throughout life. A newborn baby sleeps 20 hours out of the 24 available. This drops to about 12 hours in toddlers, 11 hours in preteens and then 9 in adolescents. An adult needs somewhere between 7 and 9 hours but this too drops to 5 or 6 hours in old age.

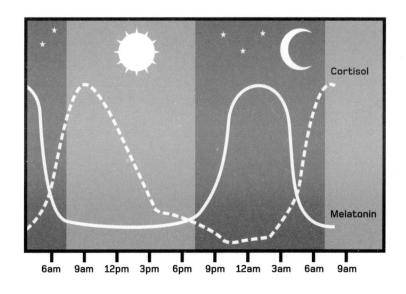

6am 9am 12pm 3pm 6pm 9pm 12am 3am 6am 9am

Cortisol

Melatonin

While levels of the sleep-inducing hormone melatonin rise in the evening, levels of the 'stress hormone' cortisol become elevated towards sunrise. This helps most people to wake naturally.

The sleep cycle

Sleep is essential. Without it, you become unable to focus your attention. Sleep gives time for the body to rest and heal, but the reason for changes in brain activity and the resulting reduction of awareness remain poorly understood.

Sleep follows a cycle that begins with a 10-minute period of reducing awareness. You are easily roused at this stage, but once you enter stage two the muscles relax and you are fully asleep. After about 30 minutes, you descend into stage three, or deep sleep, becoming more or less oblivious to outside stimuli. About 90 minutes after falling asleep, the body becomes more active again. You are in REM, or rapid eye movement, sleep. The eyes move, but the skeletal muscles are paralyzed. Dreaming occurs during REM. A dream is probably the brain performing admin, sorting through memories accrued during the day, and the mind imposes some narrative to them. You will complete about five such cycles a night. As the night progresses, periods of deep sleep become shorter.

Stages of sleep

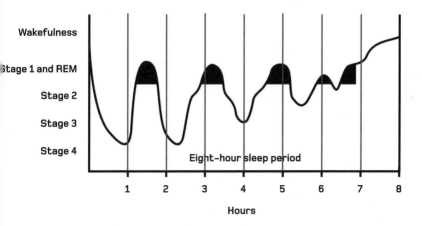

In a typical night, most people go through
four or five cycles of deep and REM sleep.

Dreams

A night's sleep involves dreaming up to five times, with the longest dream states occurring in the hours before waking. It is common for people not to remember dreams, or even be aware that they have had them. Yet others have good memories of their dreams, which feel real at the time, but have an unrealistic edge to them on waking. Some authorities claim that dreams relate to subconscious desires and concerns, while others propose we are imposing a story onto random brain activity.

During a dream, the muscles of the body are paralysed (the eyes and facial muscles can still move), perhaps to ensure we do not physically act out our dreams. We may wake from dream sleep and remain paralysed momentarily. This can be accompanied by a feeling that there is another person in the room. One explanation for this is that the brain takes its cues on body position from sensors deep in the muscles and blends that with information from the sense organs. If these two sources are not used together, the brain creates two body images for a short time.

Henry Fuseli's painting *The Nightmare* (1781) vividly captures unpleasant phenomena now known to be associated with sudden waking from a dream state, such as sleep paralysis and a feeling that others are present in the room.

Imagination

The human body has a mind, an internal entity that is able to picture, model and predict objects and events that the body cannot see or sense. This ability is closely linked with memory. Imagination is the recollection of unconnected experiences which are then associated in novel ways. In this respect, imagination is at the root of the human form of intelligence, which is based on predicting the future and then planning for it. No other animal can do this to the same extent.

Imagination also seems to have a link to language because it is played out in an internal monologue that we perceive through words: we quite literally talk to ourselves. However, it has been suggested that this imaginary monologue is overlaid on a more primitive visual system as used by other animals. Many species have been observed to play, acting out fantasy scenarios of fights or courtship. Chimps, our closet relatives, go further with games involving pretend babies and food, which suggests they may have some kind of 'inner life' of their own.

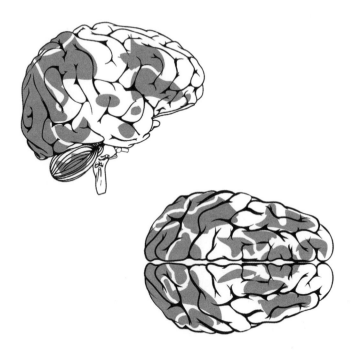

In 2013, researchers identified a widespread 'neural network' in the brain linked to visual imagination and the 'mind's eye' visualization of objects and spaces.

Theory of mind

Theory of mind is what makes every human an individual. It is the realization that the content of one's own mind – the memories, emotions and imagination – is different to that of another person's mind. The mind cannot be directly observed, however, so the idea that we all know and believe different things can only ever be an educated guess – or a theory.

The philosophical school of solipsism says that, because minds beyond our own are simply theoretical, there can be no distinction drawn between the mental and physical worlds. Therefore everything we perceive, including other people, is made within the mind. Nevertheless, the theory of mind seems to work well for the rest of us! We are not born with a theory of mind, but we figure it out at around the age of three. A game of hide and seek reveals whether a child has formed the theory yet. If they have not, they think that simply covering their eyes so they cannot see where they are will mean no one else will be able to see them either.

Very young children believe themselves—and others—to be invisible if their eyes are covered. This is true even if the rest of the body is visible.

Consciousness

'I think, therefore, I am'. French philosopher René Descartes' famous aphorism arose from his doubts about whether he was awake or dreaming. However, it was only when he was truly conscious that he would doubt whether or not he was.

Human consciousness is a nebulous concept. It is the state of awareness that involves being able to tell that the physical, objective world is distinct from the mental, subjective world, which in turn creates a sense of unique identity. Other animals have a degree of self-awareness, but whether they are conscious like humans is a hard question to answer. It is also difficult to say when humans became conscious in the past. It is debatable whether even ancient historical figures were fully conscious, because they attributed first-hand mental experiences (thoughts) to entities from the outside world – things like demons, spirits and the words of God. The brain does not appear to have a seat of consciousness, which is instead thought to arise from the association of thousands of brainwide activities.

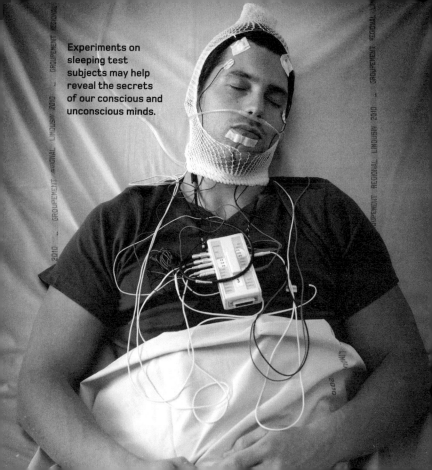

Experiments on sleeping test subjects may help reveal the secrets of our conscious and unconscious minds.

Personality

We all have a personality; it is what makes us who we are. There are many generic ways of describing personality: introverted, extroverted, generous, neurotic, narcissistic, paranoid, obsessive, open. However, these really describe behaviours, or ways of thinking, that we all do from time to time. So a personality is a constellation of traits, with some being more dominant in one person than in others – we simplify this by saying we have different personalities.

Personality is thought to develop through our experiences in early life. While the role of genetics cannot be ruled out, there has yet to be any definitive link between genes and personality. Any traits that seem to run through families could simply be the result of learned culture that is repeated by each generation. However, certain personality disorders, such as psychopathy – where somebody's way of thinking results in a danger to themselves and to others – does run in families and is likely to have a genetic component.

ENFJ Teacher	INFJ Counsellor	INTJ Mastermind	ENTJ Field Marshall
ENFP Champion	INFP Healer	INTP Architect	ENTP Inventor
ESFP Performer	ISFP Composer	ISTP Operator	ESTP Promoter
ESFJ Provider	ISFJ Protector	ISTJ Inspector	ESTJ Supervisor

Key personality types according to the popular Myers-Briggs classification. Letters indicate various character traits.

E/I: Extroversion/Introversion
S/N: Sense/Intuition
T/F: Thinking/Feeling
J/P: Judging/Perceiving

Autism

First described in 1938 by Austrian doctor Hans Asperger, autism is a developmental disorder that first becomes apparent at around the age of three. A set of similar, although less severe, disorders are also termed Asperger's syndrome.

The word 'autism' derives from the Greek word for 'alone'. Sufferers appear to shun the outside world and have difficulty interacting with others. They spend long periods performing repetitive tasks and become distressed in unfamiliar surroundings. In severe cases – the classic form of autism – sufferers cannot talk, have learning difficulties and poor motor skills. Other people on the autism spectrum may have above-average intelligence but are awkward and sometimes socially inappropriate. It appears that boys are more likely to show autistic symptoms. However, that finding may represent a gender bias in diagnosis, where autistic behaviours in little girls, such as obsessive neatness and hair brushing, are not spotted because they are not deemed particularly abnormal feminine traits.

Pervasive developmental disorders

- **Autism spectrum disorders**
  - Pervasive development disorder, not otherwise specified
  - Autistic disorder
  - Asperger's syndrome
- **Rett's disorder**
- **Childhood disintegrated disorder**

Autism and related disorders are today often grouped as 'pervasive developmental disorders' – generalized delays in the development of various functions including social and communications skills.

Human evolution

The human being, *Homo sapiens*, is a very young species, perhaps only 200,000 years old. How it arose and why is a question many humans have tried to answer.

Anatomical evidence shows that humans are in the primate class of mammals, specifically belonging to a family called the great apes. Further genetic evidence shows that our species arose first in Africa, from a common ancestor shared with

chimpanzees that lived in forests about eight million years ago. While the chimps and other extant apes are still forest creatures today, *Homo sapiens* is the last surviving species of ape that evolved to live outside of the forest.

Food and shelter are relatively easy to find and control in a dense habitat like a forest, but humans are thought to have mostly evolved in a drier savannah habitat. Out in the open, our ancestors faced a different set of challenges to secure the resources for survival. That change of habitat drove many of the adaptations seen in modern humans, with an upright stance, a large brain, language and a complex social system among them.

Our primate ancestry

Many human characteristics evolved long before our ancestors left the forests to live on the savannahs of East Africa. Some can be traced all the way back to early primates that began to evolve around 60 million years ago, and can still be seen in our distant cousins, the living primates. These include dextrous hands with nails, binocular vision that sees in colour during the day, and a decisive brain with its scheming form of intelligence.

Primates include lemurs, monkeys and gibbons. In general, they live in and around trees. To help them grip branches, they have lost most, if not all, of the hairs from their palms and soles, revealing soft fleshy pads of skin. The pads mould around branches, and any claws are replaced by nails. In addition, nearly all monkey species are diurnal (awake during the day), relying on their sense of sight to find ripe leaves and fruit. Forward-facing eyes provide binocular vision for judging distances. A larger cerebral cortex is able to map the forest in space and time, so the monkey can remember where different foods can be found with each changing season.

Lucy

In 1974, a partial skeleton and skull fragments of a human ancestor that lived 3.2 million years ago was unearthed in Ethiopia. The remains were named Lucy, because the Beatles song 'Lucy in the Sky with Diamonds' was playing as researchers discussed the find.

Lucy was an *Australopithecus* ('southern ape') and the first fossil evidence of a bidepal, human-like animal that walked upright. At just 1.1 metres (3½ ft) tall, she is considered to have been a mature female, although young, when she died. She had longer arms than a modern human, showing that she could walk on two legs, but was also adapted for climbing in trees. *Australopithecus* lived in an edge habitat between forest and savannah. Lucy's teeth suggest that she ate mainly plants. To digest this food, 60 per cent of her blood would have been diverted to the digestive system, limiting the energy available to the brain. Lucy's skull was the same size as a chimpanzee's, showing that bipedality evolved before large brains and high intelligence.

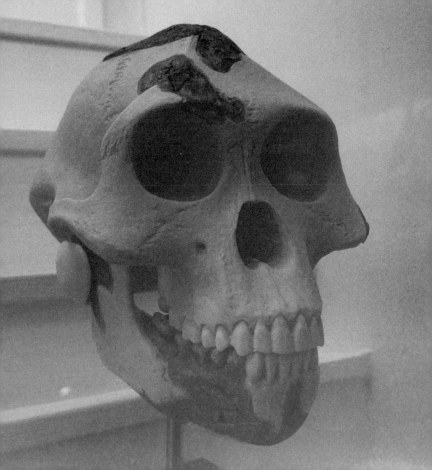

Homo genus

The modern human is the only extant member of the *Homo* genus. The first species was *Homo habilis*, or 'handy man', which appeared about 2.8 million years ago. It got that name because of the simple stone tools it made, enabling it to smash the bones of dead animals to extract the marrow. This ultra-fatty food was ideal for making large brains.

H. habilis died out 1.5 million years ago, with the rise of a larger species, *Homo erectus*, or 'upright man'. This species spread out of Africa into Asia. There is little evidence of toolmaking in the Asian population, which suggests they may have used bamboo, which would have rotted away. In Africa, this species developed a more advanced technology, using a simple toolkit to manufacture stone blades. *H. erectus* probably tamed fire as well. Finally, the African species evolved white sclera – the region of the eye around the iris seen in modern humans. It is thought this feature highlights the eyes and makes it easier to communicate using nuanced facial expressions.

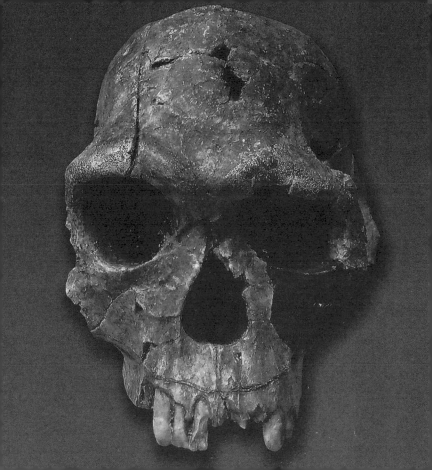

Neanderthals

Homo erectus survived in Asia until around 70,000 years ago. In Africa the species had long since given way to a series of new members of *Homo*. About 800,000 years ago, one species, *Homo heidelbergensis* arrived in Europe. By 250,000 years ago, it had evolved into the Neanderthal (*Homo neanderthalensis*).

Neanderthals are thought of as being stupid apemen, but their brains were bigger than ours, although their larynx was not adapted for speech. They thrived during the last ice age and had very thick-set bodies that were adapted to life in the cold. They hunted with staffs tipped with sharpened stone heads and killed their prey up close, rather than throwing their weapons as spears. *Homo sapiens* and *Homo neanderthalensis* lived side by side in Europe and the Middle East for about 15,000 years, and may have interbred. The last Neanderthals died about 28,000 years ago. At this time, the world was warming after an ice age, and perhaps the quiet ice people could not compete for food with their talkative, scheming neighbours.

Modern humans

The modern human, *Homo sapiens sapiens*, is a subspecies that arose in Africa from the *Homo* genus (see page 334), 45,000 years ago. This date sees the advent of art, burial, fishing and consistent use of fire – a change that is reflected in the subspecies name, which means 'wiser wise man'.

As well as bipedality, intelligence and culture, modern humans stand out from other animals in two aspects: hairlessness and a cryptic oestrus. There are two theories for the evolution of the former. One suggests that chimp-like ancestors left the forest to forage in water, swapping thick hair for a deeper layer of fat to insulate the body better. (Diving underwater required greater control of breathing, leading to an ability to talk.) It is more likely that the hair was lost to prevent overheating. Cryptic oestrus means there are no external signals that female humans are ovulating – most animals broadcast this fact. The purpose of hidden fertility is hotly debated and is linked to the complex social bonds between the sexes.

Bipedality

The benefits of standing on two legs seem obvious: it frees the hands to manipulate objects. However, it probably evolved for a combination of reasons over time.

One reason to stand up is to reach fruits among the flimsy branches of a bush. It also helps to see further when surveying the horizon for approaching danger. In addition, four-legged animals, such as antelopes, get baked by the savannah sun and have evolved ways of shedding excess heat through the nose. Our ancestors may have solved this problem by standing up so the hot sunlight had less area to heat. Most of it hit the head, which was insulated with hair and cooled with a large density of sweat glands. Once we had got up on our back legs, walking became an efficient, if slow, way of moving. We can walk for hours without getting out of breath. The hands were no longer needed for knuckle walking, but were already equipped with a firm grip for climbing. They then evolved further with an opposable thumb that provided a fine grip.

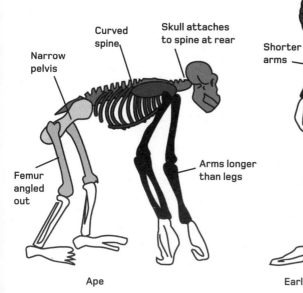

From four legs to upright walking

Narrow pelvis

Curved spine

Skull attaches to spine at rear

Femur angled out

Arms longer than legs

Ape

Skull attaches to spine below

S-shaped spine

Shorter arms

Bowl-shaped pelvis

Femur angled in

Early hominid

Brain size

Bipedality requires that the pelvis lies in the same plane as the legs, rather than being perpendicular to them, as in four-legged animals. This leaves less room for babies to pass through, so human babies are born while still small enough to fit. Human evolution probably involved a process of neoteny, where babies were born younger, after a shorter gestation. Consequently they were less developed and entirely helpless.

During gestation, the head develops first, so babies born earlier have larger brains. This anatomical shift was helped by tool use, which made it possible to access high-quality foods to build nervous tissue and provide it with energy. The human brain is by no means the largest in the animal kingdom; that of the sperm whale is five times bigger. Nor do humans have the highest brain-to-body ratio – dolphins, ants and even mice all beat us. The human form of intelligence – we must assume others exist – comes from the brain's associative powers that allow us to plan ahead based on past experience.

Relative sizes of head and body at different ages

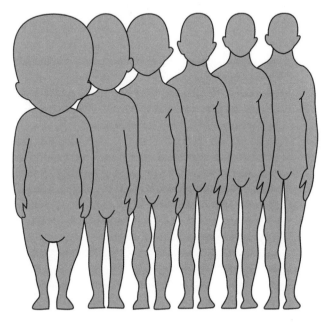

4 months   8 months   3 years   7 years   13 years   Adult

The Great Leap Forward

Evidence from human remains and the tools and goods that are associated with them suggests that, while anatomically modern humans emerged around 150,000 years ago, they did not start behaving in a modern way – the way that embodies humanity – for another 100,000 years or so.

The appearance of cave paintings, fishing equipment, ritual burial of the dead and other cultural and technological developments from 50,000 to 40,000 years ago is sometimes described as the Great Leap Forward. This suggests that our ancient ancestors had a sudden (in evolutionary terms) cognitive upgrade. Before this time, toolmaking evolved slowly and remained largely unchanged for hundreds of thousands of years at a time. However, toolmaking is governed not by genetics, but by learned culture. The Great Leap Forward is suggested to represent a genetic shift, where a new gene called FOXP2 led to advances in language. Talking allowed our ancestors to innovate, and the rest is history – literally.

The Magadelenian culture that flourished in Europe around 20,000 years ago showed new sophistication in both art and technology.

Migration from Africa

The human population is descended from a few groups, perhaps just one, that left Africa in the prehistoric past. *Homo erectus* left this territory 1.9 million years ago, spreading into Asia via a land bridge that connected Africa to the Middle East.

By the time modern humans left Africa, the only land route out was the narrow Sinai Peninsula. However, the territory beyond was dominated by Neanderthals. The speed of migration indicated that people probably left by sea, making voyages across the Red Sea and along the coast of southern Asia. They reached India 75,000 years ago and, 46,000 years ago, humans were in Australia, a land completely empty of hominids. Humans spread into Europe from Asia 43,000 years ago and were in East Asia from 30,000 years ago. The first humans walked into the Americas from Siberia 14,600 years ago. A later migration, starting 3,500 years ago, spread into the Pacific – settling in New Zealand just 750 years ago.

Major migration routes of *Homo sapiens*

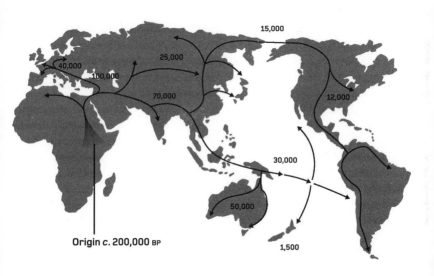

15,000

25,000

40,000

100,000

70,000

12,000

30,000

50,000

1,500

Origin *c.* 200,000 BP

Approximate dates of human settlement are shown in years before the present day (BP).

Diet and nutrition

You are what you eat. The human diet provides the energy source that powers metabolism. It contains the raw materials from which the body is built as it grows and is maintained. Food also contains vitamins and trace minerals that cannot be manufactured by the body, but are nevertheless essential to good health.

Human dentition – or the arrangement of our teeth – shows that we evolved to eat an omnivorous diet, containing plant and animal foods. We can bite and slice flesh, but also grind and pulp vegetable matter. However, it is perfectly possible to eat a healthy meat-free diet, although a vegan diet that eschews dairy products does require a supplement of vitamin B12, which is only present in animal foods.

Humans have been cooking food for at least 50,000 years, and our bodies rely on that to predigest meals before we eat them. We can no longer survive on raw, unprocessed food alone.

Human tooth types

Incisor: used for
biting food

Canine: used for ripping
and tearing food

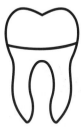

Premolar: used for grinding
and chewing

Molar: used for grinding and
chewing

Essential nutrition

A balanced diet needs to contain elements of the four major food groups: fats, proteins, carbohydrates and fibre (see page 22).

Fats are found in dairy products, meat and as oils in nuts, some fruits and other vegetables. Protein is packed into meat, the muscle of animals, but is also present in nuts and beans. Complex carbohydrates, or starches, are found in cereal products like bread and root vegetables. Simple carbohydrates are sugars that are present in sweet foods, such as honey and fruit. One carbohydrate, cellulose, is indigestible by humans. This is fibre and is in all vegetables. Fibre adds a solid element to food as it passes through the digestive tract, helping to keep it moving through the system at a healthy rate. What proportions of these food groups make a balanced diet is debatable. The human body can process most foods, storing or removing them according to its needs. Yet large quantities of refined sugar, an unnatural element, are poorly handled.

Major food groups

Vitamins

There are 13 essential vitamins required for good health. They are generally identified by the letters A to E, with eight B vitamins, including several similar, but distinct, substances. None of these chemicals can be synthesized by the body and so must form part of a balanced diet. A further chemical, vitamin K, is present in foods, but the bacteria in the bowel also produce a supply for the body to use.

Each vitamin has multiple effects on the body. They may act as hormones, as is the case with vitamin D, which has a regulatory effect on metabolism and tissue growth. Others act as antioxidants – chemical blockers that stop the oxygen entering tissues from reacting with the wrong chemicals. That would lead to the release of radicals, highly reactive chemical units that wreak further chemical damage in the cell. Vitamins A, C and E have this job. Finally, vitamins are co-factors, or helper chemicals, without which enzymes cannot perform their required function. The B vitamins are mostly co-factors.

Vitamin	Key functions include:
A (Retinol)	Maintenance of healthy eyes, bones, skin and immune system
B1 (Thiamin)	Production of ATP
B2 (Riboflavin)	Energy production
B3 (Niacin)	Carbohydrate conversion
B5 (Pantothenic acid)	Synthesis of fatty acids
B6 (Pyridoxene)	Sodium/potassium regulation, red blood cell production
B7 (Biotin)	Fatty acid synthesis
B9 (Folic acid)	Cell and DNA production
B12 (Cyanocobalamin)	DNA synthesis, maintenance of nerves and red blood cells
C (Ascorbic acid)	Synthesis of collagen, maintenance of immune system
D	Calcium and phosphorus absorption, boosting immune system
E	Antixodiant, muscle development and nerve function
K	Blood clotting and bone metabolism

Trace elements

All but a small fraction – around 1 per cent – of human tissue is made from molecules constructed from carbon, hydrogen, nitrogen, oxygen, sulphur, calcium and phosphorus. The next 0.85 per cent of the body is made up of potassium, sodium, chlorine and magnesium atoms. These are largely present as ions dissolved in body fluids to create concentration gradients across membranes – there is more on one side than the other – and that allows cells to make electrical potentials or pump materials in and out. The remaining 0.15 per cent – with a combined mass of 10 grams ($\frac{1}{3}$ oz) – is made up of 16 different elements, which only exist as a tiny trace.

Ten of the trace elements – iron, nickel, cobalt, copper, zinc, manganese, molybdenum, iodine, chromium and selenium – have known functions. For example, iron is used in the blood, while iodine is an ingredient in the thyroxine hormones. The other six trace elements – arsenic, fluorine, boron, lithium, silicon and vanadium – are always present, but their functions are unknown.

Elements in the body

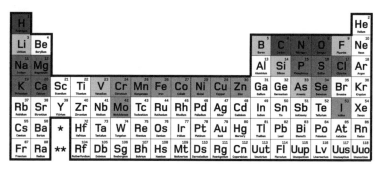

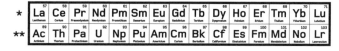

Basic organic elements

Elements present in quantity

Essential trace elements

Trace elements with unknown functions

Food calories

The calorie is a unit of energy. It is based on the amount of heat released by a reaction – most commonly something burning. This release is measured by a resulting change in temperature. One calorie of energy heats one gram of water by one degree Celsius.

The energy content of food is calculated in precisely the same way. It is placed inside a sealed chamber and burned. This replicates, by a more intense method, the way in which the body releases energy from food. However, the 'Calorie' counts listed with a capital 'C' on food packets are actually kilocalories (kcal) or 1,000 calories. One gram of fat contains approximately nine kcal (9,000 calories), while the same mass of protein and carbohydrate contain about four kcal apiece. The estimated daily energy use of a human body is 2,600 kcal in men and 2,000 kcal in women. This is a guide to how much food energy one needs to consume to maintain a constant weight. Eat less and you lose weight; eat more and the excess energy is stored as fat.

Calorie counts of some common foods

Food	Serving	kcal
Egg	100 g	155
Black tea, coffee (unsweetened)	1 cup	1
Whole milk	1 cup	148
Butter	100 g	717
Cheddar cheese	100 g	402
Sugar	100 g	387
Apple	1 large	116
Banana	100 g	121
Orange juice	1 cup	111
Broccoli	100 g	34
Carrot	100 g	41
Potato	100 g	77
Rice, white	100 g	130
Wheat flour	100 g	340
Bread (white)	1 slice	66
Bacon	100 g	540
Lean beaf	100 g	250
Chicken breast	100 g	165
Cod	100 g	82

Malnutrition

A lack of food calories is called undernutrition. This leads to weight loss, first by using up the body's fat stores. This is the process used when dieting to lose weight. As long as a low-calorie diet contains essential vitamins and nutrients, there are no impacts on health – quite the opposite, in fact, if the person is overweight. However, if the undernutrition continues for a long period, the body will show signs of malnutrition.

Without a supply from fat stores, the body is forced to convert the proteins in muscles into sugars to fuel the brain. The stomach shrinks and atrophies, so appetite dwindles, leading to dehydration that makes the skin thin and cracked. A lack of vitamins has body-wide implications, and the immune system cannot fight off infection. A lack of protein makes water flood from the body into the intestines, creating a swollen belly. Once a body has lost 30 per cent of its weight, it is classified as being in starvation. Death occurs once that loss hits 40 per cent. In most people that happens within 12 weeks without food.

Fat deposits

It is said that humans – those living in developed nations at least – have the physiology of a polar bear in the food environment of an oyster. If an oyster wants food, it simply opens its mouth. The food industry allows humans to do more or less the same. A polar bear, by contrast, spends two-thirds of the year hibernating, and so must eat as much as it can and store it as fat that adds 30 per cent to its weight. This store will fuel its body through the winter. While oysters do not lay down long-term food stores, humans do. Our ancestors would have experienced periods of feasting followed by famine. Most of us today never run out of food, but our body still prepares for the worst and as a result many of us get fatter than we need to be.

Fat is laid down in adipose tissue, a connective tissue that forms a layer under the skin, and also spreads through the body. As well as storing fat, the tissue acts as a shock absorber that protects the organs. The fat is stored as droplets inside cells called adipocytes.

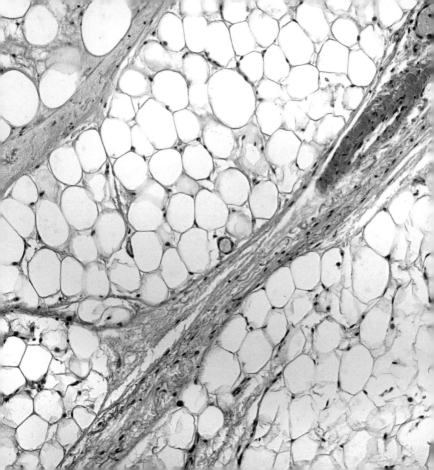

Obesity

It is estimated that 39 per cent of humans (1.9 billion people) are overweight, and 13 per cent (600 million) are obese. By contrast, only about 9 per cent of the world is underweight.

A body is regarded as overweight if it has a body mass index (BMI) above 25. (It is underweight if the BMI is below 18.5.) BMI compares a body's weight with its height. (The calculation is weight in kilograms divided by the square of the height in metres.) While the BMI is criticized for not taking build into account, it is a good indicator of health problems associated with weight. If the BMI is over 30, then the body is obese and morbidly obese at 40 and over. Obesity is the leading cause of preventable death, because obese people are very likely to suffer from type 2 diabetes or cardiovascular disease. Together these create a complex of problems called metabolic syndrome. The added weight of the body also damages joints and leads to arthritis. Obesity is caused by consistent overeating. Treatment involves dieting often with surgical interventions to limit food intake.

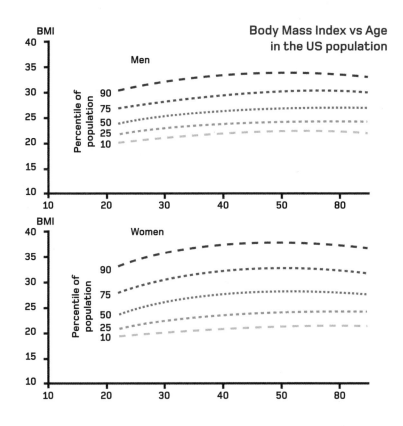

Disease and medicine

The human body is a self-regulating, self-defending and self-healing system. However, sometimes it needs a little help. Medical science aims to intervene on behalf of the body to ensure good health. To do that, doctors diagnose the problem using a series of symptoms as their clues. Treatment can be a chemical medicine that blocks or boosts a process in the body, a surgical procedure that removes or mends a problematic body part, or manipulation of the body to stimulate healing.

Medical science has its roots in ancient Greece, where disease was thought to arise from an imbalance in the body's system, which treatment then sought to redress. Modern medicine takes a similar approach, and aims to keep fluid levels, body temperature, inflammation and blood loss within safe parameters. Most diseases are self-limiting, so treatments aim to alleviate suffering while the immune system does its work. More serious diseases require more extensive interventions, where harm to the body is weighed up against the benefits of the treatment.

Major classes of medicinal drug

Class	Function	Example
Analgesic	Pain relief	paracetamol
Antibiotic	Prevents bacterial infection	penicillin
Anticoagulant	Prevents blood clots	heparin
Antidepressant	Treatment of depression	prozac
Anticancer	Treatment of cancer	cyclophosphamide
Antiepileptic	Prevents seizures	paraldehyde
Antipsychotic	Manages delusions	haloperidol
Antiviral	Prevents viral infection	sofosbuvir
Sedative	Reduces awareness and causes sleep	phenobarbital
Diuretic	Promotes urine production	furosemide

Germ theory

Through the centuries, the causes of disease have been explained in many ways. Primitive civilizations put it down to possession by demons; in ancient Greece, where the Universe was thought to be constructed wholly of fire, air, water and earth, illness was due to an imbalance of four corresponding fluids, or 'humors'; while Roman medics described disease as being spread by a miasma, or 'bad air'.

Miasma remained the leading theory until the 19th century. In 1854, John Snow provided conclusive proof that cholera was linked to unclean water. A few years later, Louis Pasteur carried out a series of experiments that showed how microorganisms, invisible to the naked eye – but largely ubiquitous – were responsible for the decay of food and communicable diseases. This discovery of germs revolutionized the role of hygiene in medical practice – and in wider society. Within decades aseptic (germ-free) conditions were used in surgery, and antiseptics and antibiotics were being developed.

Louis Pasteur introduced the technique of pasteurization to prolong the shelf-life of food. The process involved heating the food in order to kill any harmful bacteria.

Infection

Contagious diseases (the ones you 'catch') are due to a life form called a pathogen. To a pathogen, the body is a habitat within which to thrive and multiply. The reaction to that invasion creates fever as the immune system tries to get rid of it, and a host of other symptoms caused by the actions – either accidental or intentional – of the pathogen.

There are four major classes of pathogen. The largest are multicellular animals, mostly worms, which live as parasites in the digestive tract or blood supply, or embed in the organs, skin and muscle. Diseases like malaria and sleeping sickness are caused by single-celled microorganisms, such as amoebas and ciliates. These pathogens have complex cells like our own. Bacteria are the best-known pathogens, although most are harmless and indeed often beneficial to us. They have simple cell structures. Finally, most common diseases, like flu and chickenpox, are spread by viruses. A virus hijacks a human cell's own machinery to make thousands of copies of itself.

Common pathogens and associated diseases

	Description	Key human diseases
Bacteria	Single-celled organisms without a nucleus	Tuberculosis, food poisoning, staphylococcus and streptococcus infections, tetanus, pneumonia, syphilis
Viruses	DNA segments that take over living cells	Common cold, influenza, cold sores, genital herpes, AIDS, chicken pox, small pox
Fungi	Simple organisms found as single cells or thread-like filaments	Ringworm, athlete's foot, candidiasis, histoplasmosis, tinea, mushroom poisoning
Protozoa	Single-celled organisms with a nucleus	Malaria, sleeping sickness, traveller's diarrhoea, giardiasis

Viruses

A virus is a piece of parasitic DNA (or RNA, a similar chemical). They are utterly ubiquitous; a cupful of seawater contains more viruses than there are people on Earth. All viruses cause disease, but only 219 viruses are known to target humans.

Viruses cause dangerous diseases, such as HIV, zika and ebola, but also a range of less harmful illnesses, such as colds, flu and chickenpox. Viruses are very simple entities. Each is a strand of DNA held inside a protein coating. They are not alive in the accepted sense, because they do not feed, move or respond to their environment. Instead, they enter a cell and add their DNA to that of the host. The viral genes then make the cell produce vast numbers of viruses – so many that the cell eventually ruptures, releasing thousands of new viruses. When repeated on a vast scale a tissue becomes damaged, causing a disease. Their simple structure means that viruses are able to mutate very fast. This is why there is a constant threat of new forms of flu and the regular appearance of entirely new diseases.

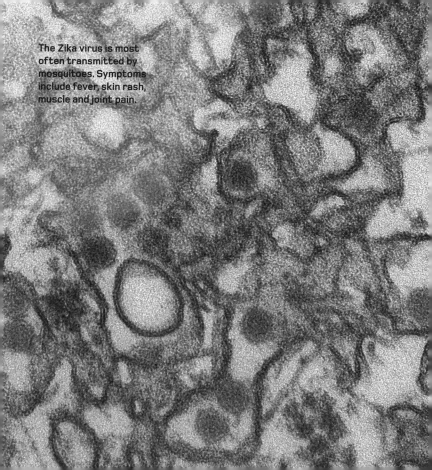

The Zika virus is most often transmitted by mosquitoes. Symptoms include fever, skin rash, muscle and joint pain.

Internal parasites

The human body can play host to many parasites. Endoparasites are the ones that live inside the body. Infestations of amoeba or protozoa cause diseases such as toxoplasmosis, leishmaniasis and Chagas' disease. However, most endoparasites are worms of various types.

Roundworms, also called nematodes, that invade the lymph nodes create elephantiasis; those that live in the skin and attack the eyes create river blindness; and pinworms are roundworms that live in the colon. They exit the anus in the evening to lay eggs, creating a feeling of fullness and an itch in the region. A quick scratch ensures the eggs are under the fingernails for eventual transfer back to your mouth. Other parasitic worms are helminths, or flatworms. These organisms have no mouth, but absorb nutrients through the skin. Tapeworms belong to this group. They use a hook-like structure on the head to hang inside the intestines, and collect nutrients from chyme as it passes. Several other flatworms, called flukes, invade the organs.

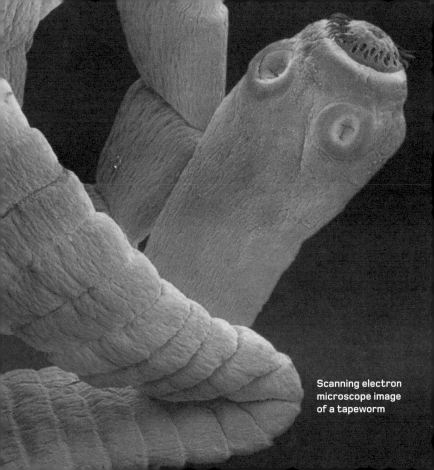

Scanning electron microscope image of a tapeworm

External parasites

In the days before washing machines and regular bathing, humans were regularly infested with ectoparasites, or creatures that lived on the outside of the body.

Scabies, an itchy skin complaint, is caused by a mite burrowing into the skin. A minute mite that lives on one-third of human faces specializes in eating the sebum produced in eyelash follicles. Ticks, relatives of mites, feed on the blood of larger animals, and will drink human blood if given the chance. The human flea is thought to have originated in South America, where it jumped species from guinea pigs. Similarly, bedbugs lurk under a mattress by day and emerge at night to drink a sleeper's blood. Obviously, bedbugs evolved before beds, and this species is thought to have been a parasite in bat roosts before shifting to our cave-dwelling ancestors. The most common ectoparasites are blood-sucking head lice, which live in thick head hair. They have two less common cousins: pubic lice live on the other body hairs, and body lice live in our clothing.

Cancer

It is estimated that somewhere between 30 and 50 per cent of humans will develop cancer at some point in their lives. It is not one disease, but several dozen that attack different organs and body systems. Cancer is a malfunction of the genes that control growth, and is more common in older people.

The prevalence of cancers today is probably due to people living to a greater age. Nevertheless, 90 per cent of them are linked to exposure to chemicals, a poor diet, radiation and infections. Tobacco causes 22 per cent of cancers, while obesity is linked to another 10 per cent and infections cause a similar number. A cancer is an uncontrolled growth of a body part. Despite the body having several lines of defence against such an event, when it happens, a mass of abnormal cells (a tumour) develops. As a tumour grows, it disrupts the functioning of organs, nerves and the blood supply. It also sheds cells that form other tumours around the body. Treatment includes removing tumours with targeted medicines, surgery or radiation that kills their cells.

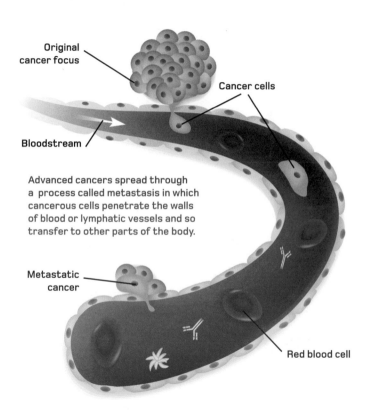

Original cancer focus

Cancer cells

Bloodstream

Advanced cancers spread through a process called metastasis in which cancerous cells penetrate the walls of blood or lymphatic vessels and so transfer to other parts of the body.

Metastatic cancer

Red blood cell

Diabetes

Around eight per cent of the world population has diabetes. The full name for the disorder is diabetes mellitus, which roughly means 'passing honey'. Sufferers produce urine containing large amounts of sugar, which ancient doctors noticed attracted bees. It is a sign that the body is not making effective use of insulin. This hormone controls blood sugar, by converting an excess of sugars into glycogen for storage.

Diabetic blood has high levels of blood sugar, which can damage the eyes and blood circulation. Blood sugar can also crash with potentially fatal consequences. Long-term cases fall into two types. Type 1 generally appears in childhood. It has a genetic component and is suspected to be linked to particular viral infections. Type 1 sufferers are unable to make their own insulin and must inject a supply throughout the day. Type 2 diabetes is due to a poor diet, which makes the body resistant to the effects of insulin. About 90 per cent of sufferers have this type. It is treated with insulin and dietary changes.

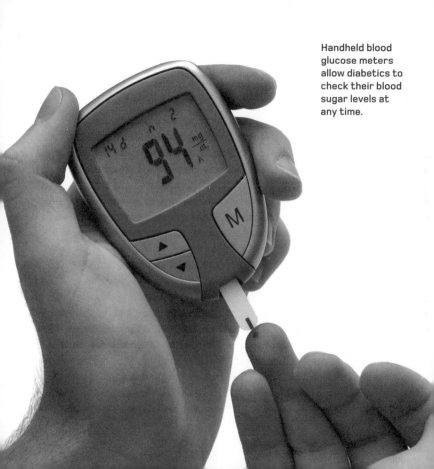

Handheld blood glucose meters allow diabetics to check their blood sugar levels at any time.

Heart disease

This mix of diseases is the leading cause of death globally, although it is less common in Africa. Better described as cardiovascular disease it is closely associated with, but not exclusive to, people with a high BMI. The disease is caused by atherosclerosis, or 'thickening of the arteries'. This is a build-up of fat in the lining of the arteries that reduces the volume of the vessels. Atherosclerosis is caused by smoking, a lack of exercise, a high-fat diet and high blood pressure. In many cases, most or all of these factors are present.

Atherosclerosis causes a number of problems, and thus kills in different ways. It can restrict the blood supply to the heart, causing heart attacks. It causes the aorta to swell and then rupture, resulting in a catastrophic reduction in blood flow. In the brain, thickened arteries stop blood reaching the brain, resulting in a stroke, where part of the brain dies. About 90 per cent of deaths from cardiovascular disease are preventable with a healthy diet and regular exercise.

Healthy artery and heart

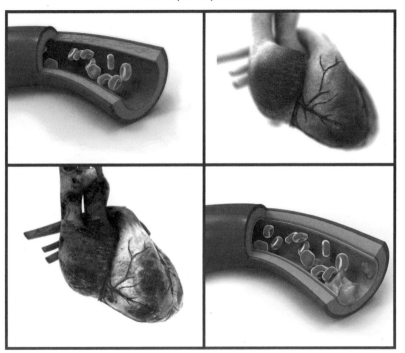

Diseased heart and artery

Geriatrics

Geriatrics is the branch of medicine that deals with diseases of old age. Chronic (long-lasting) diseases linked to ageing including cataracts (a clouding of the eye's lens), osteoporosis (weakening of the skeleton), dementia and an increase in heart problems. However, the most ubiquitous disease of old age is arthritis, a catch-all term for various disorders that affect joints.

Rheumatoid arthritis is an autoimmune problem experienced by about one per cent of people of all ages, in which the body's defence system causes inflammation in certain joints. However, osteoarthritis, where joints are damaged by general wear and tear, is far more common. The cartilage that supports a joint breaks up with age, exposing bones so they grind against each other. Pain and swelling can be controlled with drugs, but badly damaged joints (usually the hips) can be replaced surgically with metal and ceramic versions. Around 15 per cent of adults over the age of 60 suffer some form of osteoarthritis.

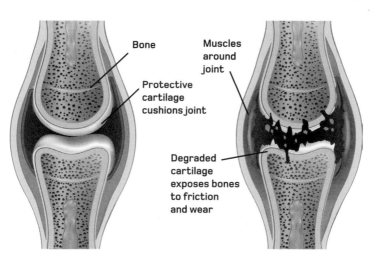

Healthy joint

Joint with osteoarthritis

Bone

Protective cartilage cushions joint

Muscles around joint

Degraded cartilage exposes bones to friction and wear

Painkillers

The medical term for a painkiller is an analgesic. These are a wide-ranging group of chemicals that stop, or at least reduce, the sensation of pain. Mild analgesics include commonly used drugs, such as aspirin and paracetamol. Stronger ones are morphine and its relative, diamorphine, more commonly known as heroin. In effect, 'recreational' users of heroin are addicted to a very powerful painkiller.

Aspirin and ibuprofen are two forms of non-steroidal anti-inflammatory drugs (NSAIDs), which reduce the sensitivity of nerves around a wound. They are good for reducing muscle pain and pain from other injuries. Paracetamol is thought to act on the central nervous system, although its action remains unknown. Morphine is an opioid – one of several drugs derived from opium, a thick sap produced by poppies. While effective, painkillers have many side effects, so they are best used with caution. Early modern surgeons, for example, made frequent use of alcohol's analgesic effects, but today it is never used in a clinical context.

Types of painkiller

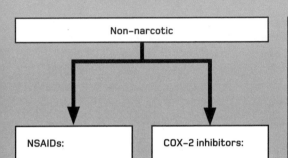

Non-narcotic

NSAIDs:

Available over the counter, NSAIDs include aspirin, ibuprofen and naproxen, generic naprosyn, generic feldene and generic motrin.

COX-2 inhibitors:

NSAIDs that relieve pain and inflammation by inhibiting COX enzyme. They include celecoxib and generic celebrex. Paracetamol is thought to inhibit COX-2 in the central nervous system.

Narcotics

Plant-based opioid painkillers and synthetic narcotics, used to alleviate pain related to the brain, central nervous system and gastrointestinal tract. Examples include morphine, codeine, oxycodone, tramodol and methadone.

Medicinal drugs

Medicines or pharmaceutical drugs are chemicals that have some therapeutic effect, either by diminishing the symptoms of a disease, by curing its cause or by preventing it in the first place. They work – if indeed we know how they work – by intervening in metabolic pathways or by upgrading the defensive powers of the immune system.

Humanity has long used substances – generally derived from plants and fungi – to treat ill health. Chimpanzees may even do it and have been observed eating specific leaves when they have an upset stomach. Those ancient remedies would have arisen by chance, tradition and guesswork. The ones that work – like chewing willow bark for pain – have persisted; this bark contains something similar to aspirin. Similarly, penicillin, the first antibiotic, was discovered by accident when fungus was seen to kill bacteria. Pharmaceutical companies still identify likely medicinal substances in nature. They then modify them into a huge range of similar chemicals, and test each one for therapeutic effects.

Mechanisms of antibiotic action

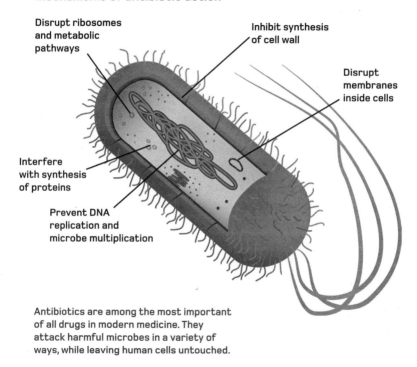

Disrupt ribosomes and metabolic pathways

Inhibit synthesis of cell wall

Disrupt membranes inside cells

Interfere with synthesis of proteins

Prevent DNA replication and microbe multiplication

Antibiotics are among the most important of all drugs in modern medicine. They attack harmful microbes in a variety of ways, while leaving human cells untouched.

Surgery

Surgeons cut the body to mend, remove or add body parts. This can range from minor procedures that do not cut into the body cavity, to major surgeries in which vital functions are performed by machines while the surgeons work on organs.

Until around 250 years ago, surgeons were not considered doctors at all. Before that, in England for example, they belonged to the Worshipful Company of Barbers. As well as cutting hair, barbers were tasked with stitching wounds and setting fractures – hence the white and red swirling sign to symbolize blood and bandages. Surgery has had a much longer history than that. It dates back to prehistory, where people had holes cut in their skulls to release evil spirits. Before the advent of anaesthetics, such as chloroform in the mid-19th century, which rendered patients insensible, surgery remained a gory and often deadly procedure. By the late 19th century, surgeons began working in aseptic (germ-free) conditions to prevent infections, while the arrival of antibiotics in the 1940s further reduced the risks.

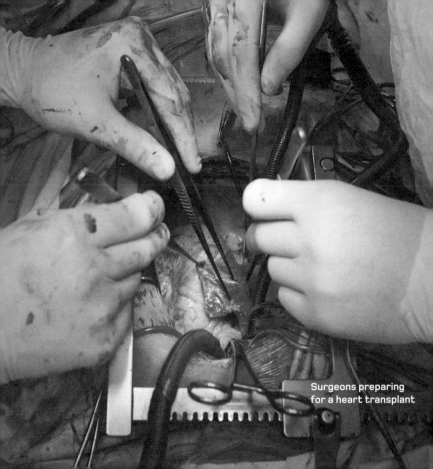

Surgeons preparing
for a heart transplant

Mental illness

A mental illness produces thoughts, feelings and behaviours that make it difficult to live a normal life. They are best understood as a disease of the brain, where the organ needs treatment just like a diseased heart or kidney would. However, because mental illness results in a shift of perceptions that can make it harder for well people to relate to sufferers, it carries a great deal more stigma.

Mental illness includes sleep disorders, compulsions and addictions, bipolar syndrome and schizophrenia. This last disorder is much misunderstood. It does not create a split personality, but makes it difficult for sufferers to differentiate between the inner and outer worlds. Treatments for grave mental illness generally involve drugs that alter brain chemistry and, in extreme cases, surgery severs brain connections. In so-called talking cures, sufferers learn to recognize unhelpful patterns of thought and use strategies to change them. Personality disorders, where people behave inappropriately, are often treated in this way.

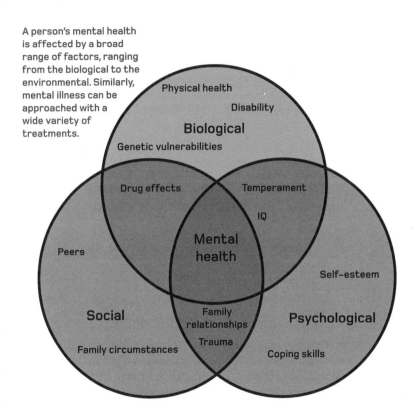

A person's mental health is affected by a broad range of factors, ranging from the biological to the environmental. Similarly, mental illness can be approached with a wide variety of treatments.

Physical health

Disability

Biological

Genetic vulnerabilities

Drug effects

Temperament

IQ

Mental health

Peers

Self-esteem

Social

Psychological

Family relationships

Trauma

Family circumstances

Coping skills

Dementia

Dementia is an incurable brain disease that results in the gradual loss of intellect, memory and sense of self, and in its final stages, the breakdown of the vital control functions of the brain, leading to muscle wastage and immobility.

The disease is associated with old age, and is predicted to become more common as the average age of the human population increases. About 3 per cent of people between 65 and 75 have dementia. This increases to 19 per cent over 75 and nearly half of all people over 85 have it. One-quarter of sufferers have vascular dementia, where cardiovascular disease has damaged the blood supply to the frontal lobes. Most of the other three-quarters have Alzheimer's disease. The cause is unknown, but results in a gradual shrinking of the brain, especially the hypothalamus, and an enlargement of the cerebral ventricles (the fluid-filled spaces within). The disease is also characterized by a build-up of plaques, or sticky masses of proteins. Most research to find a cure is focused on these features.

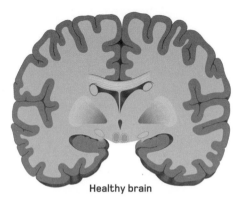

Healthy brain

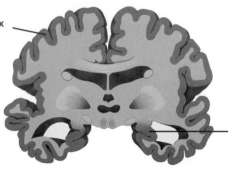

Shrinkage of
cerebral cortex

Shrinkage of
hippocampus

Brain of a person suffering from Alzheimer's disease

Ageing

The human body is able to grow from one single cell to a collection of many trillion. Once full size, it can repair all but the most serious damage. So, why can't it live forever? Instead, the body reaches a peak of fitness around the age of 30, and then gradually – ever so gradually –begins to weaken.

The process responsible is senescence, an active destruction of cells. Senescence is very much part of growth. During the early rapid development of a body, more cells are made to die off unneeded than are made to build the body itself. In later life, this same process is at work in maintaining the body. However, as the body ages, the ability to make a fresh supply of cells is reduced, while senescence continues. This process is controlled, in part, by an internal clock that allows cells to renew only a fixed number of times before they are no longer replaced. Outwardly the body ages: skin tissue becomes less flexible and wrinkles; the hair cannot make fresh pigment and turns grey. Such wear and tear cannot be fixed, but builds up until, eventually, the body dies.

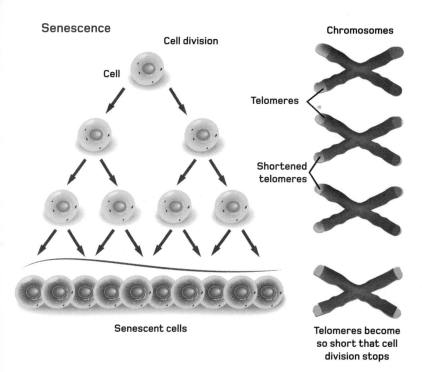

Senescence

Cell division

Cell

Chromosomes

Telomeres

Shortened telomeres

Senescent cells

Telomeres become so short that cell division stops

As we grow old, our cells stop dividing in response to a gradual shortening of the chromosomal end-units known as telomeres. Such cells do not die, but enter a state of permanent growth arrest.

Death

Death is the complete cessation of metabolism and wider body functions. Despite there being many reasons why it can occur, a human body dies in one of two ways. The first one is cardiac death, where the circulatory and respiratory systems stop, and the body is unable to receive the oxygen and fuel that are required to sustain biological processes. If the heart and breathing are not restarted, death occurs within a few minutes.

The second form of death is brain death. While the lower brain may continue to be working normally – allowing vital functions, such as heartbeat and breathing, to continue – the higher parts of the brain cease all activity. This is not so easy to confirm as cardiac death for obvious reasons. The main technique is to infuse the blood with radioactive markers. These markers travel to the regions of the body that are still living. A scan confirms brain death by showing that no markers have reached the brain.

In extreme situations, MRI scans can be used to determine whether a patient's brain is still active.

The future of the body

The theory of evolution explains how organisms change to adapt to new challenges set by the environment. Those that cannot meet them die, leaving those that can to dominate. Like no other species, the human is able to control its environment. Does that mean the human species will stop evolving? No, because we continue to adapt to combat new viruses that threaten our species. The most susceptible may perish, but a degree of immunity forms in the remaining population.

Our bodies continue to evolve owing to cultural and technological effects that alter how we live: we are taller, fatter, healthier, safer and live for longer than ever before. We are also getting more intelligent; the average IQ score has increased over the last century. Today, altering our bodies at the genetic level is regarded as immoral for all but the most extreme of cases. However, as the impacts of lifestyle accrue, our views may change. It could become acceptable to reengineer the body using genetics or implants to create a human better suited to whatever life is like in the future.

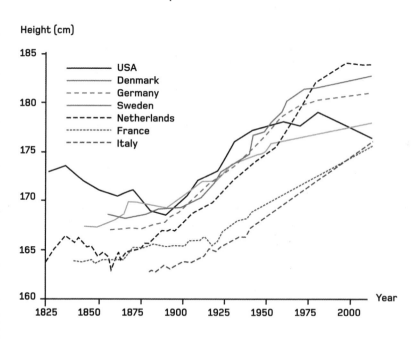

Median male height over the
past two centuries

Height (cm)

USA
Denmark
Germany
Sweden
Netherlands
France
Italy

Year

Human cloning

Cloning involves making an exact genetic copy of an individual. Mammals cannot do this naturally, but many other organisms can. Nevertheless, humans have figured out a technology that allows us to clone other organisms, including mammals.

The idea is to bypass meiosis, the process that creates sex cells (see page 256). A full set of DNA is extracted from the cell of one animal. This is then placed in an egg cell, which has had its half set of DNA removed. The egg then becomes a zygote, the first cell of a new individual. This can then develop in the normal way. However, that zygote is not really new; it is a copy, or clone, of another. Performing this process on humans is outlawed. Currently, it fails more than it succeeds, so using human subjects is morally wrong. If it could be perfected, however, would there be any point in cloning humans? Sexual reproduction works more efficiently but mixes up genes in unpredictable ways. Cloning could mass-produce humans with particular genes. Would a manufactured clone have the same rights as a sexually produced human?

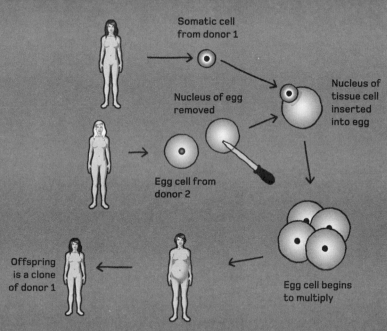

Cloning by nuclear transfer

Cyborgs

Once only found in the realms of science fiction, cyborgs – short for cybernetic organisms – are now becoming a reality. Part machine, part organic body, a cyborg has a functioning interface between the two. Cyborg technology has found most uses in medical science. The latest prosthetic limbs are controlled by detecting nervous inputs via the brain. The limb connects to a muscle near to where it joins the body. The wearer then trains the limb to move by commanding the muscle to contract – the electrical changes are picked up by the limb. Given enough time, the limb's control becomes part of the motor plans laid down in the cerebellum.

Cyborg technology has the potential not simply to fix or replace malfunctioning body parts, but also to enhance them. Artificial retinas could help us see better – even add mixed reality layers to our vision. Already researchers are using implants to send nerve signals through the Internet to control objects. Perhaps one day this technology could connect two nervous systems.

Immortality

'**D**eath is nothing to us, since when we are, death has not come, and when death has come, we are not.' So said Epicurus, an ancient Greek philosopher who questioned the need to fear death. Nevertheless, the prospect of dying is met with dread by most of us. Is it possible to live forever?

In essence, there are two ways to die: a violent event that destroys the body, or a series of medical insults that weaken it until it can no longer function. Medicine and healthy living can already slow this descent into decrepitude, and future technology may be able to stop the ageing process altogether, while intensive care medicine could support life while otherwise-fatal conditions are tackled. The result would be a body capable of living forever (barring accidents). But does immortality require the body to be maintained in perpetuity? An immortal human need only preserve its identity, which comes from mind, memory and consciousness. Perhaps the most effective route to eternal life, then, is to decode the contents of the brain and replicate it in a machine.

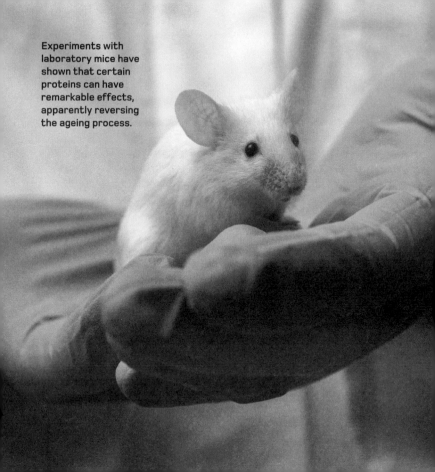

Experiments with laboratory mice have shown that certain proteins can have remarkable effects, apparently reversing the ageing process.

Mind or body?

What is it to be human? Is it the mind or the body that makes us what we are? René Descartes is the leading figure in dualism, which states that the mind is separate from the body. He said the ethereal, incorporeal mind communicated with the body by making the pineal body quiver and twitch. Our understanding of neuroscience gives us a different view.

The mind is a construct of the brain, but we have the sense that it oversees the body, exercising free will over its actions. However, in experiments, the brain appears to ready itself for action (commanding a movement) before we make the decision to do so. This fraction of a second suggests one of three things. Firstly, our consciousness is slightly delayed but is nevertheless a true reflection of our mental processes. Secondly, consciousness is a bogus narrative that we fit to an entirely subconscious control system. Thirdly, we do not exercise free will, but 'free won't'. Our decisions are made subconsciously, but the consciousness has a right of veto over them. Which one works for you?

Glossary

Allele
A version of a gene; for example the gene for eye colour has several alleles, giving rise to blue, brown, grey, hazel and green colours.

Amino acid
An organic compound containing nitrogen; amino acids are the subunits that form the building blocks of proteins, and as such form a major component of the human body. While some amino acids can be synthesized in the body, others must be obtained from diet.

Antibody
A chemical 'tag' attached by the immune system to a foreign organism that is invading the body. Antibodies may directly neutralize the invader, or otherwise mark it for attack by other parts of the immune system.

Artery
A blood vessel that carries oxygen-rich blood away from the heart.

Cell
The basic structural unit of the body, a self-replicating chemical factory consisting of various molecules in a cytoplasm surrounded by a cell wall.

Chromosome
A complex structure containing genetic information in the form of a long, coiled strand of DNA encoding many separate genes.

Cilia
A hair-like extension from a cell membrane, used to create currents in fluid surrounding the cell.

Collagen
A structural protein used in many body parts, such as skin and ligaments.

Corpuscles
Another word for a cell, or collection of a few cells – generally one that stands out in form and function from the surrounding tissue.

Cytoplasm
The liquid contents of a cell.

Diploid
Descriptive of a typical body cell that contains two complete sets of chromosomes. Most cells in a human body are diploid, the only exception being sex cells used in reproduction.

Distal
The furthest point from the core of the body.

DNA
A complex 'nucleic acid' molecule that is used in all living organisms to carry reproducible genetic code – instructions for the manufacture of proteins that allow the body to develop and function.

Embryo
An unborn offspring in the early stages of its development.

Endoplasmic reticulum
A network of membrane surfaces and tubes within the cytoplasm, where many of the cell's components are manufactured from raw materials.

Enzyme
A protein that is involved in metabolism by controlling a specific chemical reaction needed for life.

Evolution
The mechanism by which humans change over many generations, as environmental pressures act on random genetic variations, ensuring that some traits spread through a population while others disappear.

Flagellum A tail-like propulsion unit found on many cells.

Foetus
A developing baby that is more advanced in its form than an embryo but is still unable to survive outside the mother.

Gene
The basic unit of genetic inheritance. The gene can be regarded as a

strand of DNA that codes for a particular protein, or it can be seen as a distinct characteristic inherited from a parent, such as hair colour.

Genome
The full collection of genetic material in a species, including genes and 'non-coding' sections of DNA.

Genotype
A description of the alleles carried by an organism.

Glycogen
A complex and highly branched carbohydrate molecule similar to starch, used to store energy-rich sugars in human cells.

Glycolysis
The metabolic pathway that forms the first stage of cellular respiration, in which energy is released from glucose.

Golgi apparatus
An organelle that is used to package materials for release, or secretion, from cells.

Haploid
Describing a cell that contains only a single set of genes. In humans, only sex cells (the female eggs and male sperm) are haploid.

Hormone
A chemical messenger that is released by glands into the blood stream and which may have multiple effects on different body parts.

Metabolism
The series of chemical reactions by which a living organism extracts energy from food or other fuel sources, creates the chemical building blocks needed for its growth and reproduction, and eliminates harmful waste.

Organ
A collection of tissues in the body that are joined together to perform a specific function. Most organs are self-contained units, although the skin is also generally regarded as an organ.

Organelle
A machinelike structure in a cell that performs a particular set of functions.

Osmotic pressure
The force that pushes water in and out of cells, between areas of different chemical concentration.

pH
A measure of acidity on a scale from 0 to 14. Substances with a pH of less than 7 are acidic, while those with pH above 7 are alkaline.

Phenotype
A description of the physical and behavioural traits of an organism – the visible outcome of its genotype.

Plasma
The liquid fraction of blood.

Protein
A complex molecule used by all living things to build structural body parts such as muscle, and also as enzymes.

Puberty
The period of development during which a child becomes a sexually mature adult.

Respiration
The process that takes place in every living cell to extract energy from a food source, such as sugars.

Sphincter
A ring of muscles that opens and closes vessels and tracts in the body.

Stimulus
Something that is detected by a sense organ and passed onto the central nervous system.

Tissue
A collection of cells with similar properties and a shared function in the body, often as part of an organ.

Tract
A tube or vessel within the body that ultimately connects to the outside world.

Vein
A blood vessel that carries deoxygenated blood from body organs and muscles back towards the heart.

Virus
A non-living disease-causing agent based on DNA that takes over the machinery of a cell.

Zygote
The first cell of a living body, formed by the fusion of two haploid sex cells.

Index

action potential 144–5
adolescence 280
 female 274–5
 male 276–7
adrenal glands 220–1,
 222, 226
adrenalin 220, 222
Africa, migration from
 346–7
ageing 394–5, 382–3, 404
alimentary canal 64, 86
alleles 292
allergies 136–7
alveoli 92, 93
Alzheimer's disease
 392, 393
anaphylaxis 136
anti-diuretic hormone
 (ADH) 224, 232
antibodies 128–9, 130,
 298
arm 50, 51, 60, 61
arteries 102, 103, 114,
 116–17
arthritis 382

Asperger's syndrome
 326
autism 326–7
autonomic nervous
 system 178–9
axons 142, 144, 145, 148

balance, sense of 172–3
behaviour, human
 302–26
bipedality 332, 340–1,
 342
bladder 236–7
blood 102, 104–8
 clotting of 124–5
 types 298–9
bodily secretions
 202–22
body mass index (BMI)
 362, 363
body systems 28–9
bone 34–5
bone marrow 108–9
brain 152–6, 306, 307,
 322, 406

brain death 396
brain size 342–3
breasts 278–9
breathing 94–5
breathing rate 224,
 228–9

calories, food 356–7
cancer 376–7
capillaries 102, 114,
 120–1
carbohydrates 22, 350
cardiovascular system
 102–12, 114
cartilage 36–7
cell division 256–7
cells 16–17, 26, 394
cerebellum 154,
 158–9
cerebrum 154, 160–1
chemicals for life 22–3
chromosomes 290–1,
 296
circadian rhythm
 310–11

circulatory system 28, 29, 114–20
cloning, human 400–1
cochlea 170, 171, 172
collagen 36, 38
colon 76, 82–3
conception 254–5
consciousness 322–3, 406
contraception 258–9
copulation 252–3, 254, 258
cortisol 220, 226
coughing 96–7
cranial nerves 162–3
cyborgs 402–3

death 396–7, 404
defecation 86–7
dementia 392–3
dermis 180, 181, 184–5
diabetes 378–9
diet 348–62
digestive system 28, 29, 64–86
DNA (deoxyribonucleic acid) 16, 284, 286–7, 288, 290
dreams/dreaming 314, 316

dualism 406

ear 168–72, 206
ectopic pregnancy 262–3
egg, human (ovum) 244–5, 246, 254
endocrine system 202, 204, 214
epidermis 180, 181, 182–3
essential trace elements 14, 15
evolution, human 328–46, 398
eye 162, 164–6, 192, 208
eyebrows 192
eyelashes 192

fat deposits 360–1
fats 22, 350
fibre 350
fight-or-flight response 220, 222–3
foot 54–5
future of the body 398–406

gall bladder 74–5, 81
gas exchange 92–3, 94
gender 282–3

genetics/genes 284–8, 292, 300, 324
genotype 292–3
germ theory 366
glands 202, 204–5
goblet cells 210, 211
Great Leap Forward 344
growth 272–3

hair 188–9
hair colour 190–1
hand 52–3
hearing 170–1
heart 110–11, 114
heart disease 380–1
heartbeat 112–13
hiccups 100
homeostasis 224–36
Homo erectus 334, 336, 346
Homo genus 334–6
Homo sapiens 328–9, 336
hormones 156, 202, 212–13, 214, 216, 218, 220
humours 366
hunger 66–7
hypodermis 180, 181, 184–5, 194

hypothalamus 156, 218, 230, 232, 310

imagination 318–19
immortality 404–5
immune system 122–38
immunity 130–1, 132
infection 368–9
inflammation 134
inheritance 284–300
insulin 214, 224

joints 30, 40–1

kidneys 232, 234–5

language 306–7, 344
leg 50
ligaments 38
limbic system 156, 174
limbs 50–1
liver 80–1
Lucy 332–3
lungs 88, 90–1
lymphatic system 138–9

malnutrition 358–9
mammary glands 278–9
man, average 12–13

mandible 42, 43
medicinal drugs 386–7
medicine 364–96
menstrual cycle 246, 248–9, 274
mental illness 390–1
metabolism 18–19
mind 320–1, 406
modern humans 338
mucus 204, 210–11
muscles 29, 30, 31, 38, 56–61, 150, 158
    antagonistic pairings 60–1
    contraction 58–9

nails 186–7
nature vs nurture 300
Neanderthals 336, 346
negative feedback 226–7, 232
nervous system 29, 140–78
neurons 140–1, 142–3, 152
nose 174–5
nutrients, absorption of 78–9
nutrition, essential 350–1

obesity 362–3, 376, 382
organs 26–7
osmoregulation 232–3
ovaries 238, 240, 244
ovulation 246–7

pain 200–1
painkillers 384–5
pancreas 204, 214–15
parasites
    external 374–5
    internal 368, 372–3
parasympathetic system 178, 179
pathogens 368, 369
pelvis 48–9
peristalsis 84–5, 86
personality 324–5
phenotype 294–5
pituitary gland 218–19, 224
placenta 260–1
plasma 104, 138
pregnancy 254, 260–70, 278
    first trimester 264–5
    labour and birth 270–1
    second trimester 266–7

third trimester 268–9
primates 330
proteins 22, 350
puberty 174–6, 280

red blood cells 104, 106–7, 108, 109
reflexes 150–1
reproductive system 238–82
respiration 20–1
respiratory system 28, 29, 88–100
ribcage 46–7
running 62

saliva 70–1
satiety 66–7
sebaceous glands 204, 206
sebum 204, 206–7
sex determination 296–7
sex organs
  female 238, 240–1
  male 238, 242–3
sexual intercourse 252–3, 254, 258
sexuality 280–1

skeleton 28, 29, 30, 32–3
skin 26, 180–200, 206
skull 42–3
sleeping 312–16
small intestine 74, 76–7, 78–9
smell 174, 176
sneezing 98–9
social groups 304–5
somatic nervous system 146–7
somatosensory cortex 198–9
sperm 242, 244, 250–1, 252, 254
spinal cord 148–9, 150, 162
spine 44–5
spot formation 207
stem cells 108, 256
stomach 72–3
surgery 388–9
sweating 194–5, 230
sympathetic nervous system 178, 179

taste 176
tears 208–9
teeth 68–9, 348, 349

temperature control 230–1
tendons 38, 39
thorax 46–7
thyroid gland 216–17, 226
tissues 24–5, 26
tongue 176–7
touch, sense of 196–7
trace elements 354–5

umbilical cord 260, 261
urea 234

vaccinations 132–3
veins 102, 103, 114, 116, 118–19
viruses 368, 369, 370–1
vision 166–7
vitamins 352–3
vocal cords 308–9

walking 62, 63
white blood cells 108, 109, 122, 126–7, 130
woman, average 10–11
worms 368, 372

zika virus 370, 371

First published in Great Britain in 2017 by

Quercus Editions Ltd
Carmelite House
50 Victoria Embankment
London EC4Y 0DZ

An Hachette UK company

Copyright © Quercus Editions Ltd 2017
Text by Tom Jackson
Edited by Anna Southgate
Packaged by Pikaia Imaging

A CIP catalogue record for this book is available from the British Library

PB ISBN 978 1 78648 123 8
EBOOK ISBN 978 1 78648 124 5

Every effort has been made to contact copyright holders. However, the publishers will be glad to rectify in future editions any inadvertent omissions brought to their attention.

The picture credits constitute an extension to this copyright notice.

10 9 8 7 6 5 4 3 2

Printed and bound in China

Picture credits 2: GraphicsRF; 11: Oguz Aral; 13: Oguz Aral; 15: Martial Red; 17: Andrea Danti; 25: Christopher Meade; 27: bluezace; 29: GraphicsRF; 31: stihii; 33: stihii; 35: Yoko Design; 37: Blamb; 39: Sebastian Kaulitzki; 41: Blamb; 43: eveleen; 45: Peter Hermes Furian; 47: corbac40; 49: Alila Medical Media; 51: Linda Bucklin; 53: corbac40; 55: BlueRingMedia; 57: Designua; 59: Alila Medical Media; 61: stihii; 63: Michael D Brown; 65: La Gorda; 67: Designua; 69: graphicgeoff; 71: Alila Medical Media; 73: Alexander_P; 75: joshya; 77: Tefi; 78–9: Tefi; 81: MSSA; 83: Tefi; 85: Designua; 87: Blamb; 89: Alila Medical Media; 91: Alila Medical Media/Andrea Danti; 93: Andrea Danti; 95: eveleen; 103: chombosan; 105: Designua; 107: royaltystockphoto.com; 109: Alila Medical Media; 111: Nicolas Primola; 113: Alila Medical Media; 115: NelaR; 117: Designua; 119: Designua; 121: mmutlu; 123: NoPainNoGain; 125: Timonina; 127: Alila Medical Media; 129: Designua; 131: Kateryna Kon; 135: Suzanne Tucker; 137: Designua; 139: Alila Medical Media; 143: Tefi; 145 top: Tefi; 147: stihii; 149: Alila Medical Media; 151: NoPainNoGain; 153: Jesada Sabai; 155: Tefi; 157: decade3d - anatomy online; 159: decade3d - anatomy online; 161: Christos Georghiou; 163: ellepigrafica; 165: Blamb; 167: Alila Medical Media; 169: Sedova Elena; 171: Alexilusmedical; 173: Alexilusmedical; 175: Yoko Design; 177: joshya; 179: Alila Medical Media; 181: BlueRingMedia; 183: komkrit Preechachanwate; 185: AkeSak; 187: Yoko Design; 189: Designua; 191: Alila Medical Media; 193: Sergey Peterman; 195: Andrea Danti; 197: Designua; 199: OpenStax Collegevia Wikimedia; 201: Blamb; 203: Jamilia Marini; 205: Designua; 207: Designua; 209: Alila Medical Media; 213: Designua; 215: Alila Medical Media; 217: Tefi; 219: Designua; 221: Designua; 223: BOV; 225: Image Point Fr; 231: Blamb; 235: Blamb; 237: Blamb; 239: Gagliardilmages; 241: Designua; 243: La Gorda, 245: Sebastian Kaulitzki; 247: Tefi; 249: Tefi; 251: Songtum Prakobtieng; 255: Alila Medical Media; 259: vadim-design; 261: Alila Medical Media; 263: yaruna; 265t: Lemunik b: La Gorda; 267t: Lemunik b: La Gorda; 269t: Lemunik b: La Gorda; 271: Alila Medical Media; 273: Moremar; 275: Peter Gardiner/Science Photo Library; 277: Peter Gardiner/Science Photo Library; 279: Blue Ring Media; 285: India Picture; 287: Calmara; 289: Designua; 291: Zuzanae; 294–5: Rawpixel.com; 299: Designua; 301: wong sze yuen; 303: Anastasia Mdvanian; 305: Mert Toker; 307: okili77; 309: stockshoppe; 319: Christos Georghiou; 321: Peter Polak; 323: Burger/Phanie/Science Photo Library; 331: Anna Kucherova; 333: Ramavia Wikimedia; 335: Locutus Borgvia Wikimedia; 337: Tim Evanson/Flickr; 339: zatvornik; 345: World Imagngvia Wikimedia; 349: eveleen; 351tl: Joe Gougherstorkcerstock inc.; tcl: multiart/; tcr: Evlakhov Valeriy/; tr: Nattik/a; bl: Somchai Som/; bcl: Nattika/; bcr: WitR/; br: Valentyn Volkov/; 359: Jenny Matthews / Alamy Stock Photo; 361: Jose Luis Calvo; 371: CDC/ Cynthia Goldsmithvia Wikimedia; 373: Power & Syred/Science Photo Library; 375: Protasov AN; 377: Designua; 379: Alexander Raths; 381: Giovanni Cancemi; 383: Tefi; 387: Designua; 389: Dmitry Kalinovsky; 393: Alexilusmedical; 395: Designua; 397: Allison Herreid; 403: Philippe Psaila/Science Photo Library; 405: Mirko Sobotta; 407: Rafael Ramirez Lee. All pictures obtained via Shutterstock, Inc. unless otherwise indicated.
All other illustrations by Tim Brown.